T0313848

Experimenting
in Tongues

WRITING SCIENCE

EDITORS Timothy Lenoir and Hans Ulrich Gumbrecht

Experimenting in Tongues

STUDIES IN SCIENCE
AND LANGUAGE

EDITED BY *Matthias Dörries*

STANFORD UNIVERSITY PRESS
STANFORD, CALIFORNIA
2002

Stanford University Press
Stanford, California

Printed and bound by CPI Group (UK) Ltd,
Croydon, CR0 4YY

Library of Congress Cataloging-in-Publication Data

Dörries, Matthias
   Experimenting in tongues : studies in science and language /
edited by Matthias Dörries
      p.   cm. — (Writing Science)
   Includes bibliographical references and index.
   ISBN 0-8047-4441-6 (alk. paper) —
   ISBN 0-8047-4442-4 (pbk. : alk. paper)
   1. Science—Language.   2. Communication in science.
I. Dörries, Matthias.   II. Series.
Q226 .E97   2002
501'.4—dc21   2001057652

Original Printing 2002

Last figure below indicates year of this printing:
11   10   09   08   07   06   05   04   03   02

Typeset by James P. Brommer in 10/14 Sabon
and Helvetica Black display

CONTENTS

CONTENTS

MATTHIAS DÖRRIES is professor of the history of science at the Université Louis Pasteur in Strasbourg and a research fellow at the Max Planck Institute for the History of Science in Berlin. He is the coeditor of *Restaging Coulomb* (Florence, 1994) (with Christine Blondel) and *Heinrich Kayser: Erinnerungen aus meinem Leben* (Munich, 1996) (with Klaus Hentschel) and has written articles about the interface between the natural sciences and the human sciences. Currently he is finishing a book with the title *The Future of Science in Nineteenth-Century France (1830–1871)*.

ANN GENEVA received her Ph.D. from the State University of New York at Stony Brook in 1988. Her book *Astrology and the Seventeenth Century Mind: William Lilly and the Language of the Stars* was published by Manchester University Press in 1995.

EVELYN FOX KELLER received her Ph.D. in theoretical physics at Harvard University, worked for a number of years at the interface of physics and biology, and is now professor of history and philosophy of science in the Program in Science, Technology and Society at MIT. She is the author of *A Feeling for the Organism: The Life and Work of Barbara McClintock*; *Reflections on Gender and Science*; *Secrets of Life, Secrets of Death: Essays on Language, Gender and Science*; *Refiguring Life: Metaphors of Twentieth Century Biology*; and *The Century of the Gene* (Harvard University Press, 2000). Her most recent book, *Making Sense of Life: Explaining Biological Development with Models, Metaphors, and Machines*, is due to appear in the spring of 2002 from Harvard University Press.

CHRISTIAN LICOPPE (CNET/CNRS), a physicist and historian of science and technology, currently heads a laboratory at France Telecom R&D in Paris that studies the application of the social sciences to new information and communication technologies. He is the author of a book on proofs in experimental accounts in the modern period, *La formation de la pratique scientifique* (Paris: La Découverte, 1996).

JÖRG PFLÜGER studied mathematics and philosophy and received a Ph.D. and a habilitation in theoretical computer science. Currently he is a professor at the Vienna University of Technology, in the Institute for Design and Assessment of Technology, part of the computer science department. His research interests concern the epistemology and culture of computer science and media technology.

ROBERT J. RICHARDS is professor in the departments of history, philosophy, and psychology at the University of Chicago. He is the author of *Darwin and the Emergence of Evolutionary Theories of Mind and Behavior* (University of Chicago Press, 1987) and *The Meaning of Evolution: The Morphological Construction and Ideological Recon-struction of Darwin's Theory* (University of Chicago Press, 1992). He has a book in press bearing the title *The Romantic Conception of Life: Science and Philosophy in the Age of Goethe*.

STEPHANIE SUHR is a biologist and linguist from the University of Freiburg. This chapter derives from her thesis, "Ist der Begriff der Sprache auf die Gene übertragbar?"

While studying the work and life of the German spectroscopist Heinrich Kayser, I came across Arnold Sommerfeld's epochal book *Atombau und Spektrallinien* of 1919. In the preface, Sommerfeld asserted that the problem of the atom would be solved once "the language of spectra is understood"— a physical goal phrased in philological or linguistic terms. Although clearly metaphorical, the quotation nevertheless raised my curiosity about how language and the study of language actively shape scientific research. This question seemed all the more pressing to me, since the language metaphor has become a commonplace in discourses of and about genetics with the Human Genome Project, which aims to "read" the "book of life" and "decode" the "genetic language." My curiosity resulted in a conference, "Language as an Analogy in the Natural Sciences," which took place at the Forschungsinstitut für Technik- und Wissenschaftsgeschichte of the Deutsches Museum in Munich in November 1997. The conference brought together some sixteen scholars from various fields, analyzing how computer scientists, biologists, mathematicians, naturalists, astrologers, chemists, and physicists have looked at nature and their own work through the lens of language.

This volume presents a selection of the contributions to this conference. Obviously, it can have no pretensions even remotely to cover an immense field, but it claims that knowledge of and experiments with languages informed and continue to inform inquiries into nature and therefore actively shape our world and culture. The aim is to bring together some elements of the larger historical picture, to raise key issues, and to provide brief overviews of more recent developments, with the hope that these issues will be taken up more thoroughly in the future.

The final editing of this volume took place during my stay at the Max Planck Institute for the History of Science, Berlin, and the beginning of my new work at the Université Louis Pasteur in Strasbourg. The Deutsche Forschungsgemeinschaft generously provided funding for the conference. I would like to thank the Forschungsinstitut für Technik- und Wissenschaftsgeschichte and especially its director, Helmuth Trischler, for strong support during the organization of the conference and the editing of the book. The library of the Max Planck Institute for the History of Science has been extremely helpful and efficient in providing relevant literature during various stages of the editing of this book. Many thanks also to Sven Marcus Kleine, who digitally prepared the illustrations. The volume has equally profited from commentaries by three anonymous referees and from the attention of skillful editor Nathan MacBrien of Stanford University Press. Finally, I would like to thank the contributors to this volume and all the participants at the conference for many stimulating discussions.

Matthias Dörries
Berlin, February 2001

Experimenting
in Tongues

CHAPTER ONE

Language as a Tool in the Sciences

Matthias Dörries

> Die Sprache ist ein Instrument. Ihre Begriffe sind Instrumente. . . .
> Begriffe leiten uns zu Untersuchungen. Sind der Ausdruck unseres
> Interesses, und lenken unser Interesse.
>
> Language is an instrument. Its concepts are instruments. . . .
> Concepts guide us to investigations. They are the expression of
> our interest, and they direct our interest.
>
> —Ludwig Wittgenstein, *Philosophical Investigations*

References to language abound in the sciences: biologists speak about *reading* the human genome and *rewriting* the genetic code; computer scientists talk about *programming languages*, and mathematicians about a *universal symbolic language*. What is behind these references to language? What do they say about how science actually works? This book aims to retrieve from a variety of perspectives the historical, methodological, and ideological motivations that encourage the use of the language metaphor. Do analogies to language have an orienting function at the beginning of exciting new research, or are they useful over longer periods? Can they be replaced by other more fruitful analogies? Does nature ever actually work and develop the way language does?

Asking such questions means arguing that language matters for scientific investigation; it implies that language has a certain generative power in science. Language is not just a means of expression or the medium through which scientists preferentially argue their cases. It can also serve a quite different function as a conceptual tool for scientific inquiry; and scientists have indeed self-consciously instrumentalized their means of communication, as the chapters in this volume show. What then have scientists' options and choices been in referring to language? There is no straightforward answer. Rather, choices change over time with ever-evolving knowledge about language and equally with shifting interests and means of inquiring into nature. We may perhaps speak of mutual adaptation between the linguistic and scientific realms and of continuous adjustments between knowledge about language and knowledge about nature. Tracing and analyzing the scattered elements of this history is what this collection is about.

It may be useful at this point to remark on the relations between the approach chosen here and the recent wave of rhetorical studies of science. In the preface to *The Literary Structure of Scientific Argument* (1991), editor Peter Dear spoke of language as "a shaper (perhaps a realizer) of thought and an embodiment of social relations."[1] Although Dear still suggested studies both on language as an active force in scientific investigation *and* language as a means of persuasion, the emphasis has clearly shifted to the study of language as a *literary technology* in the quest for social recognition and power.[2] One perhaps unintended but tacitly accepted consequence was the delegation of *language* to a subsidiary role. When "language is little more than a subordinated expression of the social conditions from which it emerged," then analysis of language and rhetoric simply served to unmask scientists' agendas.[3] Foucault's deliberately vague and all-inclusive notion of "discourse" set the stage for a wide range of studies on the exercise of power via language, on how we are "governed and paralysed by language."[4] Parallel to these developments, studies in semiotics à la Roland Barthes explored other means of persuasion beside the written and spoken word, extending the notion of language to systems of signs, such as pictorial or bodily languages. This semiotic turn, however, came at the price of continual dilution of the notion of *language*.

I am not satisfied with the implicit outcome of most rhetorical studies: the reduction of language to a mere indicator of the social forces involved. I do not believe that language can become an infinitely malleable instrument of persuasion in the hands of scientists in their quest for power and authority. Neither do I wish to work here with an all-encompassing notion of language as used in semiotic studies. Rather, I would like to suggest that language (understood as either spoken, written, or formal languages that can be written down and form a history of texts) is a product of both nature and nurture. Language is a result of human evolution as well as a social convention developed over time and, for that very reason, creates and constrains the way we acquire knowledge. Therefore, language does not necessarily assume the transparent role that rhetorical analysis frequently likes to attribute to it.

Often enough it is rather opaque and bears the traces of its past. One way to take language's tenacity into account could be to look more closely at the long tradition of rhetoric, the art of persuasion. Not only does rhetoric have an epistemological dimension, but it also has an aesthetic one: "a style argues."[5] A closer look at rhetoric may be helpful to explain why certain metaphors—although obviously inadequate—are so difficult to shake. Another way to take language's tenacity into account is chosen here: rather than studying language as a medium of power, this book proposes to focus on language as a tool to generate knowledge.[6]

Language's power as a conceptual resource derives from its specific position between culture and nature. Inasmuch as language is natural and a product of evolution, some of its patterns mirror similar characteristics found elsewhere in nature. At the same time, because language is culture as well, it can never be completely brought into a one-to-one relationship with nature. Here lies the force of the analogy, because complete identity kills metaphor. Language as an analogy operates at a distance to the objects of natural research, reflecting them only inadequately; as a result, there is further enrichment in the act of comparison, and language may generate knowledge and shape scientists' ideas by providing a wider range of possible analogies. Inasmuch as language is a social, historically grown convention, serving communication, it provides a model for forms of interaction in nature. Yet again, it operates at some distance from nature: communication between human beings cannot be completely identified with communication between non-human beings. This tension, too, generates knowledge by inviting natural scientists to draw analogies. The chapters in this volume bring to light this productive side.

METAPHOR, ANALOGY, MODEL

My use of the terms *metaphor, analogy,* and *model* is pragmatic: I look to their overlap as transfers of ideas and concepts. This may be unacceptable to philosophers and linguists, but it is hard to see that the effort of drawing finer distinctions would be reasonably proportional to the result in a historical and epistemological study of the kind intended here. Rather, I will limit myself to a few useful distinctions and terms.

The first point concerns epistemological range and dynamics. Analogies can be merely illustrative; they can take on a heuristic function over a longer period; or they can be constitutive, that is, fulfilling the task of scientific explanation.[7] In fact, scientists have consistently pushed new metaphors to their limits, systematically exploring their potential to become constitutive. This means extending the scope of an analogy from obvious comparisons to less obvious but potentially even more exciting ones. It does not mean that the metaphor needs to be correct in all or even most of its aspects, since often there are no better or appropriate alternatives. Furthermore, metaphors may undergo significant modifications within new fields of use and finally become constitutive after a period of assimilation. The tendency toward rendering metaphors constitutive is not a one-way street; the reverse has also taken place, as Ann Geneva demonstrates in this volume for seventeenth-century astrology.

A second point concerns the spectrum of metaphors and analogies. The chapters in this volume contribute to a history of references to the cryptic language of stars, or genes, or to the formal languages of instruments, mathematics, and programs—whether written by God, evolution, or human beings. But beyond these literal and explicit analogies there is a much wider set of secondary metaphors, such as the *book of nature, letters, transcriptions, translation, grammar, copies, readings, libraries, transcriptions, punctuation,* and so forth. These satellite metaphors highlight certain aspects over others. Their emergence, disappearance, and clusters are useful indicators as well of language's role in the sciences.

The third aspect pertains to the effect of metaphors, which serve as mediators, or "vehicles," as Evelyn Fox Keller calls them. They transfer knowledge among distant areas and among author and audience; they provide the means for the interactive process between language and action. Thus they create as well as challenge our analytical classifications, divisions of labor, and disciplinary boundaries by suggesting unfamiliar similarities. Depending on historical circumstances, metaphors can experience periods of favor and disfavor. Taming the use of metaphors stood programmatically at the top of the agenda of a group of new seventeenth-century experimentalists, who opted for analytical clearness and the soberness of scientific language. They regarded the transfers performed by metaphors to be deceptive rather than revealing, as undesirable interference with scientific work. Specific historical circumstances may encourage or discourage metaphoric transfers, and the analogy with language may generate exciting research as often as it misleads.

CONCEPTUAL DIVISION

Although I look at analogies historically, for references to language in science I would nevertheless suggest a threefold division, distinguishing among references to the *history*, the *structure*, and the *practice* of language. The first gives priority to a diachronic view, starting from historical developments, as discussed in fields like comparative philology or historical linguistics. The second privileges a synchronic approach, analyzing a language as it operates at a given point in time and emphasizing its lawful regularities and patterns. The third looks at language from a pragmatic point of view, stressing its communicative aspect; it analyzes explicit uses of language-as-communication either for explaining apparently similar exchanges in natural phenomena or for modeling the interfaces between human and machine in computer science. In some ways, this division also reflects changing priorities in linguistic studies during the last two hundred years, from comparative historical linguistics to structuralism to pragmatics.

Analogy to the History of Language

During the nineteenth century, geologists and biologists turned to comparative philology and historical linguistics to explain the pattern of nature's development, as Robert Richards's chapter in this volume demonstrates. Charles Lyell's book *The Geological Evidence of the Antiquity of Man* (1863) included a chapter on the "Origin and Development of Languages and Species Compared." Charles Darwin used the genealogical relationship between languages to illustrate the descent and natural classification of species; he even compared unpronounced letters in the spelling of a word to rudimentary organs in animals. Inversely, philologists and linguists looked to Darwin for the confirmation of comparative and genealogical studies of languages. As Lyell's title already implied, the hope was to explain human origins by means of the origin of language and vice versa; this—at least in Lyell's case—in a way consistent with Christian thought. While Lyell tried to combine an ahistorical creationist account with a historical view of the emergence of human beings and language, Darwin's evolutionary theory went all the way and argued for parallel patterns of descent, conceiving from an apparent isomorphism a direct relationship.

What then made Darwin's analogies so forceful and convincing? Part of the answer lies with the history and emancipation of biology and philology, each of which developed solid disciplinary structures and methods in the first half of the nineteenth century. Only with two self-confident and independently working disciplines, each existing in its own right, could the analogy be more than mere chance or speculation and actually lead to mutual reinforcement.

In fact the analogy came to the forefront at a specific historical moment when, although linguists and scientists worked separately, they still talked to each other constantly. They met not only in salon and at professorial meetings but also on vacation. Furthermore, most natural scientists still had a thorough training in languages, ranging from Greek and Latin to sometimes Sanskrit or even Egyptian hieroglyphics. The most prominent and visible interaction of this exchange was in the work of the Humboldt brothers. Historians have only recently revived this so far more or less vanished world and its impact on research during the nineteenth century.[8] Before 1800 there are examples of single savants who actually worked in both fields: Thomas Young, who successfully studied both optics and the deciphering of Egyptian hieroglyphics; Joseph Priestley, who gave *Lectures on the Theory of Language and Universal Grammar* before turning to chemistry. In the first half of the twentieth century these close links appear weakened or lost. In the 1930s Max Planck phrased his scientific work merely metaphorically in the terms of philological work when he wrote, referring to Helmholtz, that quantitative data in the natural sciences were "forever only uncertain messages or . . . signs, which the real world transmits [to the scientist] and from which he then tries to draw conclusions, like the philologist who has to decipher a document from a completely unknown culture."[9] The public meetings between the biologist François Jacob and the linguist Roman Jakobson in the 1960s that Stephanie Suhr refers to in her chapter seem extraordinary, required much organization, and had lost much of the self-evident, normal character they might once have had. Exchanges then seem to have happened either on a much more abstract level, or in new emerging disciplines, like the fashionable computer sciences whose developments—as Jörg Pflüger's chapter shows—followed those in linguistics surprisingly closely.

Robert Richards's study on Schleicher and Darwin in this volume thus gives us a specific moment particularly suited to the analogy. The natural scientists he discusses still had a humanistic education with an emphasis on classical languages, which facilitated transfers of language metaphors. And these transfers had an ideal location in the German university, which brought together all disciplines under one roof, as in Jena, where Ernst Haeckel explained Darwin to August Schleicher.

Darwin's path toward the full-fledged analogy was twisted. Darwin easily adopted whatever metaphor or analogy came to hand, and the linguistic metaphor and language history was originally just one among many.[10] It had potential and seemed promising; however, up to 1859 the idea floated in Darwin's and Wilhelm von Humboldt's minds without ever being developed systematically. Then came a moment of dramatization after the publication of the *Origin of Species*. Taking up Darwin's rather casual remarks in the *Origin of Species*, Lyell brought the analogy from the merely illustrative to a heuristic level, attracted by the possibility of filling in fossil gaps by looking at the more complete historical records in linguistics. Although clearly unhappy with Lyell's obscure primary causes (which left natural selection as only a secondary cause), Darwin could live with language as a heuristic analogy, as it did not threaten his theory. However, by the end of the 1860s Alfred Russel Wallace's arguments against natural selection forced him to push the analogy further, in order to find other ways than spiritual forces to explain the refinement and perfection of human intelligence. Now language provided the appropriate tool to invalidate Wallace's claims. It became for Darwin the motor of intellectual evolution by a feedback loop of inherited alteration, an idea directly counter to British philosophical empiricism, where language could only reflect, not create ideas or intellect. Darwin argued that language might mold or create human mind (with such an acquisition becoming a permanent, hereditary legacy), drawing—as Richards shows—on the ideas of the German linguist Schleicher and thus relying on a tradition of German romantic idealism, especially Wilhelm von Humboldt. In this way Darwin elevated language from a subordinate position and merely heuristic function to an active cause, as Alexander von Humboldt had also done under his brother's influence. For the naturalist, language was not only the mirror of nature, but "reacts at the same time upon thought, and animates it, . . . with the breath of life."[11]

Three things seem striking in this case. First, Darwin developed the analogy in a defensive mood and only when he urgently needed an argument against his opponents. Second, in his response to Wallace, Darwin made use of the theory of the inheritance of acquired characters, which now appears an unnecessary assumption. Third, as Richards argues, Darwin wished to explain human progress, the large brain of human beings, in a way different from his creationist or spiritualist opponents. As we now conceive it, the assumption of progress was equally unnecessary to the core of Darwin's theory; natural selection and the struggle for existence could have provided sufficient explanation. Surprisingly, then, unproven and unnecessary assumptions need not lead to worthless analogies. Rather, the language analogy ultimately survived some of the theoretical conditions that brought it to the forefront.

While Darwin refrained from positively arguing his case, Schleicher eagerly absorbed the analogy and incorporated it systematically into a coherent theory of the history of languages. The history of languages was for him the main feature of the development of human beings; being the exclusive property of humans, it gave the highest criterion for classifying them. His embrace of monism provided a theoretical framework, which crystallized in the notion of the "language organism," parallel to the natural organism. Language, then, was the material side of mind, and the formation of language was comparable to the evolution of the brain and the organs of speech. Schleicher's reception of Darwin was therefore enthusiastic, because he found independent confirmation of his own fully worked out scheme. Schleicher pushed the analogy to its limits, which in the end even forced him to drop his initial belief in the decline of languages as incompatible with Darwin's ideas. Here we have an example of how the use of an analogy led to transformations in the field to which it was applied. Schleicher was much more concerned than Darwin with making his theory consistent with the analogy.

The case of Darwin and Schleicher show that the same analogy can mean very different things to different people. There is no *one* analogy, agreed on by the actors. Not only do analogies become important in specific moments of research, sometimes lying years apart, but their understanding and use also depend on the theoretical framework. The theoretical framework defined the power of the analogy: for Schleicher it was strictly constitutive, whereas for Darwin it only had a high potential to be so. Schleicher made his system depend on the analogy in order to preserve overall coherence, whereas Darwin simply added it as yet another confirmation of a preexisting framework. Thus the value of the analogy was different to both: for Schleicher a *single* analogy to human descent served as confirmation for his thesis, while for Darwin the analogy was only one among many, another confirmation of his overall argument. Also Darwin's tendency to push the analogies to the limit was therefore much less developed. In the end it seems for Darwin the value of a single analogy was of rather limited use, unless it fit into a whole set of analogies.

One last point: analogies can have lasting legacies. In this case they survived surprisingly in a visual form, embodied in the drawings of pedigrees (*Stammbäume*). As Richards shows, language trees, which Schleicher used systematically from the 1860s, have come to provide the chief model for the representation of descent. Here then the two disciplines, which already operated at distance, came to find common ground.

Analogy to the Structure of Language

Accounts of language emphasize its double feature: its order and disorder. The Janus face of language with its regularities and irregularities was displayed in the contrast of the Adamic Godlike language with those after the Tower of Babel; in the opposition of analogists and anomalists in ancient Greek culture; in Wilhelm von Humboldt's distinction between inner and outer discourse; finally in Saussure's opposition of *langue* and *parole*. Both the orderly and disorderly aspects have provided plenty of analogies for understanding natural phenomena, independently of whether language was regarded as belonging to nature or convention. There was a third option: not only was it possible to map orderly nature onto orderly language, or unorderly nature onto unorderly language, but one could also—as did the Stoics—start out from an orderly nature and assume that language with all its irregularities did not adequately mirror it.

Darwin succeeded in giving a persuasive account of both order and disorder in nature and language by introducing the element of chance. However, emphasis had long lain exclusively on the orderly aspects of language: on the search for a long-lost original, paradisiacal language, or for a new universal, abstract, and quantitative language that would fit nature's order and permit communication among naturalists. The former tried to express the harmony of the universe via an elaborated, correctly spoken or written language, while the latter focused on describing singular natural phenomena as succinctly as possible in a minimal number of words and with the help of mathematics. Both attempts share the assumption of a unique, God-given language, through which natural phenomena can be adequately and coherently expressed.

Both emphases are also deeply embedded in the cultural and religious traditions of the Western world. Two lines of Judeo-Christian tradition, closely intertwined, inform the historical importance of the central analogy and over the course of history set into the universe of science all kinds of satellite metaphors. On the one hand is the authority of the written word, of the book (whether Torah or Bible), and with it the belief in laws written by God; on the other hand is God's creation of nature by the act of speaking. Here the Old and New Testaments diverged on the significance accorded to language. While Genesis sees language as a tool for a God who performs through language, the Book of John claims at the beginning that language (the word) *is* God: "In the beginning was the Word, and the Word was with God, and the Word was God." Given the identity of the word (*ho logos*) with God, nature now *is* God's expression, following the rules of his authoritative language. Unsurprisingly, then, understanding nature meant finding the phrasing and the terms given by God to nature, retranslating nature into language—or to discover the code in which it was written, to decipher the divine seals created by God to conceal the true architecture of the universe.[12] Here lies a significant difference from Genesis's understanding of language as merely God's chief tool, which by itself does not necessarily provide direct access to nature.

The most prominent offspring of the language analogy was certainly the idea of the *book of nature*, this worldly complement to the essential book, holy scripture. Augustine incorporated the pagan idea of a "descriptio divina," an enciphered nature, into Christian thought, and during the Middle Ages the term could either be metaphorical or provide a framework for interpreting nature.[13] In early modern times exegesis of the book of nature reached equal status with its textual counterpart in the argument that God revealed himself through both books. Galileo was perhaps the most prominent opponent of the Averroistic separation of knowledge and belief and objected to pointless debates about words. He made use of the metaphor in his most famous statement:

> Philosophy is written in this grand book—I mean the universe—which stands
> continually open to our gaze, but it cannot be understood unless one first
> learns to comprehend the language and interpret the characters in which it
> is written. It is written in the language of mathematics, and its characters are
> triangles, circles, and other geometrical figures, without which it is humanly
> impossible to understand a single word of it; without these, one is wandering
> about in a dark labyrinth.[14]

For Galileo, God no longer expressed himself in a spoken language, but in a "lingua matematica." The quotation emphasized this other meaning of the term "logos" (as used by John): "calculation." In fact, the term *logos* was a multipurpose word in Greek, ranging from "word" (in our current meaning), to "reason," "computation," and "accounting" (among many other possibilities).[15] The order or laws handed down by God could therefore have a textual structure, as well as a much more abstract one, and why not a mathematical or geometrical one? Nature was a text, or at least was understood to be a text, either the story of a natural history (*historia naturalis*), where the abundance of language might literally represent the opulent wonders of the natural world, or a script that required interpretation, translation, and decoding.

Astrologers of early modern Europe wished to translate the divinely ordered world into human language by decoding God's "starry language" as it appeared in the heavens. As Ann Geneva shows, the analogy between language and nature operated within a Neoplatonic framework, within which one could explain one set of things by another, believing in a "perfectly ordered universe in which each element reflected and signified all others, structured by Pythagorean notion of musical harmonies, the medieval doctrine of signs, and the idea of a natural hierarchy of creation with the heavens as the apex."[16] This theoretical framework rendered the analogy between nature and language constitutive. Virtually anything could be linked to anything in most mysterious ways: for example, letters of an alphabet could be read out of the configuration of the stars. With the analogy having an explanatory function, the problem was to diagnose and translate the language and the grammar of the heavens in order to understand God's prophecies and the universe. This meant—as Geneva stresses—reaching beyond the accidental or even essential properties of things "to divine what these things signified beyond themselves."

This analogy started to break down with the Scientific Revolution, once belief in a self-signifying language was shaken. There was no longer hope of overcoming language's disorder, and a more reliable and fruitful alternative seemed to lie in the conception of a more formal, transparent universal language, which would be superior to common language with its multiple denotations. As Slaughter has pointed out, "Science provides a view which implies that language reflects properties of a real world rather than of the mind." The motivation behind it "was more scientific than linguistic"; concern lay rather "with nature than with language."[17] Language lost its privilege as the sole means to truth, replaced in the new experimentalism by new standards of shared consent expressed in sober language. While language remained central to natural history, hopes of finding an original, divine language were increasingly supplanted by detailed, accurate description. The demise of the constitutive analogy to language happened slowly; as Geneva points out, a single person could use various layers of analogies that might range from constitutive to metaphoric for different purposes.[18]

Post-World War II genetics brought the structure analogy back to the spotlight, as discussed in Stephanie Suhr's chapter. The four nucleotides—adenine, thymine, guanine, and cytosine—were now part of the "letters" of a "genetic alphabet" (A, T, G, C) and form the genetic three-letter words, the codons, which encode or "signify" one amino acid. Now it seemed possible to assume an analogy between two relationships: between cells and bodies, between sounds and word-significants. The solution seemed to lie with structural elements shared by both language and genes, such as linearity, hierarchy, redundancy, and context-dependence. Developments in cybernetics, information theory, and computer technology, together with molecular genetics and linguistics, brought these analogies center stage from the 1950s.

On the linguistic side, Noam Chomsky's hierarchical and generative principles made use of Ferdinand de Saussure's ideas about language's inner form, its patterns and structures, and located these in the brain. Chomsky's renaturalization of language, together with his quantitative approach, appealed to those biologists working on the double helix and the deciphering of the genetic code during the 1950s and 1960s. The intersection of cybernetics, structuralism, and molecular genetics led to formal analogies previously ignored in the historicist perspective Richards describes. The order came no longer a priori from God, but now lay within nature itself, in its own self-organization, a principle at the origin of both life and language. Divine creation was replaced by nature in feedback mode. The analogy was drawn most clearly by the linguist Roman Jakobson, who argued that "language was, in its architecture, modeled on the principles of molecular genetics, because it is as much a biological phenomenon as this structure of language."[19] The theoretical framework was the speculative assumption of a common blueprint for human beings and language. How human language's structure could resemble heredity's remained a question for which the biologist François Jacob also had no answer, speculating on that the analogy could derive from similar functions.

Two things seem striking. First, the central question was how out of a limited number of elements there could arise an unlimited number of meanings, variety, and something so complex as life. This question pertained to language (built on sound), genes (built out of nucleotides), as well as computer science (working with a digital code).[20] The vague notion of *information* seems to have played an integrating role. It thus posed the problem of the origins of life and language neither from a historical framework, nor by referring to a creator; rather, it tried to solve the issue in a playful, combinatoric manner, composed out of law and chance, leading to dynamic states of order. Although these discourses did not evoke God, they still aimed at an equivalent coherent, explanatory framework. Such an all-encompassing worldview avidly adopted any possible analogy at hand—at some cost, as Suhr shows.

Second, although the metaphor does not hold very far, it is nevertheless omnipresent—in scientific as well as popular discourses. Suhr brings out its limits and the numerous inconsistencies in its uses and points also to the vicious circles it creates as definitions of language depend on definitions of life and vice versa. Only at the price of a reductionistic notion of language, which excludes language as a mental affair, can we assume some similarities, which of course are then true of all systems that transmit information. Evidently analogies need not necessarily be unequivocal in order to survive, but why should this one in particular have such a powerful legacy? Perhaps simply because there is no better one, although this is hard to prove. More important, however, is the observation that language functions as the least common denominator in other interdisciplinary discourses too (such as cybernetics or structuralism), as well as in popular and religious discourses (in order to convince governments and attract public attention).

A radical step in the opposite direction was to give up the search for a given, original order in language and to create new languages that not only perfectly fit natural order but also provided a tool for prediction and research. This would mean abandoning passive reception for intervention and action. The creation of a new and superior language, replacing common human language with its imprecisions and irregularities, would be the philosopher's stone, solve all scientific questions, and allow swift and precise communication and dissemination. Already in the fifteenth and sixteenth centuries the revival of the classical languages had rendered uniform the complex jargon of sixteenth-century miners, as Marco Beretta has argued in his study on Agricola.[21] During the following centuries scientists conceived numerous new languages, mostly drawing on familiar spoken or written ones. Indeed, many conventional languages derived from an analogy to Latin. Zdravko Radman has pointed to the important issue of neologisms, the coinage of words, as a variety of metaphor. "A radical innovation must be accompanied by adequate lexical invention, for only this sort of invention can faithfully denote something that does not resemble anything familiar." Radman speaks of a "metaphoric hybridization of meaning" where "new meaning is achieved through the fabrication of old linguistic material."[22]

The denigration of vernacular languages and jargon also stood at the center of the seventeenth century's universal language movement, which condemned words as obstacles to be overcome not by rediscovering an original self-signifying language, but by creating a syntactical taxonomy. For these reformers, the astrological language, which had occupied "the middle ground between the abstract symbolic system of classical mathematics and common discourse, as it did between divine and human language," was too confusing.[23] The universalists opted for austerity and purity over variety. This could either mean turning toward quantification and mathematics, or, with Boyle, searching for new ways of using symbols to adequately express chemical phenomena and concepts.[24]

In the eighteenth century, language became the starting point for public enlightenment, as political and social advance depended on its progress. The "action of language served as a master key, disclosing the foundations not only of the separate projects of philosophizing, teaching and governing, but also of the relations joining these projects into an interdependent whole."[25] For the adepts of radical language reform, a "language could be a 'faithful mirror' of nature only through artifice, the conventional manipulation of arbitrary signs."[26] The new language's strength would lie in the disconnection between the word and the thing, thus in its arbitrariness, as realized most prominently by Linnaeus in botany.[27] In fact, arbitrariness was the only guarantee of objectivity and disinterest, surmounting the disturbances of tradition and vernacular understanding. Truth lay in an exact, expressive language, as Condillac argued: "The art of reasoning is in truth only a well constructed language."[28] Language would make nature's order self-evident.

Chemistry was a major testing field for these ideas. Lavoisier and others stressed the predictive power of an appropriate chemical language: "Languages are intended, not only to express by signs . . . the ideas and images of the mind; but are also analytical methods, by means of which, we advance from the known to the unknown, and to a certain degree in the manner of mathematicians."[29] Nomenclature and scientific practice went hand in hand. Language was neither only the rhetorical means of persuasion, nor merely the convention used by scientists to convey their experimental data; it was something more: an "epistemological conception of science that connects facts and experiments into a coherent scientific theory."[30]

Another line of attack lay in departing rigorously from the analogy to human language and conceiving highly abstract and symbolic systems together with an algorithm of operation, as with the development of an algebraic calculus. This "langue de calcul" (Condillac's term) functioned yet again as a tool for research and also as a means for communication, at least among adepts.[31] For a natural philosopher like Newton the problem remained how to translate problems from verbal language to algebraic language, and inversely how to interpret algebraic results.

In the first half of the twentieth century the ideal of a pure scientific language was taken to extremes by the logical positivists, who aimed at a language that was both literal and logical—thus the opposite of ordinary languages and metaphors with their multiple signification. The paradox of such a program of "linguistic hygienics" became obvious.[32] A "Science without linguistic concepts would collapse into an a priori mathematics unrelated to the physical world."[33] Indeed, physicists, for example, have insisted on the centrality of metaphors for their work. For the British physicist James C. Maxwell they were not only "legitimate products of science, but capable of generating science in its turn."[34] The German physicist Werner Heisenberg stressed the prophylactic impossibility of a program of a "cleansing of language from all unclear terms before all science": "The demand to define terms already in advance would be more or less synonymous with a requirement to anticipate the whole future development of science by logical analysis."[35]

Analogy to the Practice of Language

Scientists could also look to the practice of language, to human communication, in order to draw analogies in their work. When the eighteenth-century French savant Réaumur referred to a "language of a thermometer," the metaphor pointed to the interaction or comparability of different types of thermometers, and—as Christian Licoppe argues in this volume—finally to the interaction (intersubjectivity) between the savants themselves. It seems odd to attribute a language to dead objects and assume communication between them; indeed, Réaumur distanced himself explicitly from the metaphor by adding a qualifying "so to speak." There was no deeper theoretical reflection behind this utterance. Réaumur was concerned neither with the history nor with the order of language, but rather with establishing common standards, to overcome the de-localization of action and share experience at a distance. This required the adaptation of instruments, new means of registering data, and also the subversion of ordinary language.

Here, the analogy to language clearly lost in scope, figuring only as one element among others for doing science. The orderly aspect of language was present, but not essential. Consequently any analogy drawn to language would also be more modest and of a more illustrative character. The seventeenth- and eighteenth-century seekers after a new and universal language had aimed to provide both a tool for research and a means of communication, giving priority to one or the other according to their preferences. However, the more science became a collective enterprise, the more language lost its function as the master key to nature. Comparative standards were now the aim, to be rendered possible through devising a common language. Communication itself was an indispensable precondition for achieving any results at all, but not the ultimate aim. The creation of a new language became inseparably linked with experimental practice, an entanglement that Licoppe sees as the pillar of social order and a sign of modernity.

The analogy of language-as-communication came to full bloom after World War II as information theory was developed in such publications as Norbert Wiener's *Cybernetics*, in computer technology, and in the pragmatic turn of linguistics, which led to the "golden age of the linguistic metaphor in genetics" (Keller). The chapters by Evelyn Fox Keller, Jörg Pflüger, and Stephanie Suhr all show how closely ideas in computer science and information theory responded to current paradigms in linguistics, tracing their development from structural considerations toward communication theory and interactive models. In the biology of the 1950s and 1960s the diffuse term *information* became the center point of analogies between linguistics and the life sciences. Exchange of information corresponded to *communication* within the cell. Suhr shows, however, that the term *communication* can only be metaphorical, having little to do with actual human communication, since, for example, the receiver and sender, having different biological compositions and being qualitatively different, cannot exchange places. Therefore, Suhr suggests that it makes more sense to compare the genetic code with artificial languages as developed in computer programs.

Keller takes an equally critical stance about the appropriateness of linguistic metaphors for vital processes, looking at uses of the term *translation* in genetics. Paradoxically, the German biologist August Weismann used this term to describe not a symmetrical two-way communication, but rather the solely one-way connection (according to him) between germ cells and somatic cells. Some fifty years later Francis Crick identified DNA as the "text" of life, and his "Central Dogma" argued for a one-way flow of information within the cell, an idea that continued to be promoted under the heading "translation." But molecular biologists of the 1960s not only claimed to "read the text, but immediately speculated about how to "rewrite" it, thus now actually proposing to "translate" back and forth.

Keller reflects on the power of language metaphors: they are not restricted to the cognitive function of describing natural objects (as with germ cells and somatic cells, or DNA and proteins), but also function as messages to audiences: other biologists, or politicians, investors, and the public at large. In this way they order the world and affect the larger social order. In fact, there exists an interactive process between language and action, in which metaphors are vehicles for carrying the effects of our experience as material actors into language and back again. On the one hand, Keller claims that the language metaphors of the 1960s, which spoke of rewriting the gene, ultimately were partly put into realization; on the other, the realization of this program, as for example in the Human Genome Project, has affected the one-way-flow-structure of the reading paradigm. Biologists' attention shifted to the conditions necessary to make cells read genes; they turned the cell into a highly coordinated system, a vast internet, with multiple exchanges of messages. Keller draws the conclusion that there is no such thing as a passive reader, neither in the cell nor in our world, because the activities in the cell only transform messages into messages, and we as readers have the potential to transform the meaning of what we read.

Three observations ensue. First, inconsistent analogies have a life of their own, and they may even gain in favor once the central paradigms within a discipline have changed, notwithstanding that even then they may not be fully appropriate. Second, metaphors have no intrinsic value, but are only as good and valuable as we want them to be. In this sense, metaphors are like mirrors that society and the natural sciences hold up to themselves, providing images of the journey undertaken. At the same time, their presence is not only the result of social, political, and scientific forces but also affects the way we conceive and frame future action. Here we have, as in Licoppe's case, an entanglement of metaphors with meaning and practice. Third, if what has been said so far applies to any metaphor, what then is specific to the analogy of language-as-communication? This particular type of language metaphor refers not only to communication among natural objects but also to the *act of communication itself*. These language metaphors point to themselves as vehicles of communication (what Keller describes as "built-in reflexivity"). This impales us on a dilemma: the analogy to language-as-communication can only be made within language. This means, however, that there is no absolute fixed point of reference, which could fully validate the analogy, hence the diminished role of language as one factor among many that enters into doing science.

Computer science reflects on a technological level this epistemological shift toward pragmatics and communication. Rejecting structural similarities between formal and human languages, Jörg Pflüger sees language as a living process, inseparable from its uses. Although computer languages seem at first sight to be purely mathematical formalizations, they are, as Pflüger argues, much more: they are media that capture images of the world and mediate back and forth between human beings and artifacts.

Both human language and general linguistic and philosophical conceptions have served as models for programming languages and understandings of human interaction with the computer. At the very beginnings of computer science, the term *coding schemes* was more prevalent; the term *language* was used rather in the sense of a symbolic mathematical language. But more abstract problem-oriented languages, which delegated the actual coding to the computer, gave programmers more time to focus on modeling real-world problems. Programmers then suggested that user-oriented programming languages developed from natural languages, which would allow "conversation" with the computer. Pflüger shows how this idea of conversation depended on a strong belief in formal semantics, grammar, and the possibility of automatic translation from verbal communication about goals to a formal specification and program. According to this line of thought, programming languages worked as abbreviations of other languages, not as independent conceptual units or as guides of thought; the computer was degraded to a mere automaton. This, of course, required the programmer to conceive the whole program in advance, which—given the more and more difficult tasks handed to programmers—turned out to be an impossible task.[36]

The metaphors of translation and language misled programmers, Pflüger claims, who during the 1960s and 1970s turned to an alternative framework.

Now metaphors of construction and building came into use. Interactivity between the human and the computer was no longer conceived in terms of *conversation*, but now as the *manipulation of pictures*. From the 1980s, finally, in the era of the Web, the emphasis has turned again to the analogy of language-as-communication. In programming languages guided by conceptions of communication and performative action, language is no longer limited to the descriptive function of representing action in a formal model, but rather "the discursive activity of modeling itself is somehow expressed in the medium of the programming language," as Pflüger phrases it. These languages model open systems with roughly equivalent human and artificial agents that can act autonomously and interact with one another and their environment. They represent themselves and cooperate, with both becoming more and more entwined and interdependent. Here the boundaries of human-computer communication and human-human communication are increasingly blurred. Pflüger describes the overall development of programming languages as reaching toward subjectivity, away from earlier formalisms.

A striking aspect in Pflüger's chapter is his emphasis on the way the latest programming languages work in collaboration with their users in a feedback mode: apparently endlessly malleable, flexible, and adaptable to our needs, but also determining our way of life. We formalize "living" mediation processes, but the formal languages we develop affect our minds, may furnish a framework for rational behavior, and serve as a "grammar of action," to use the phrase introduced by Philip E. Agre. Wilhelm von Humboldt's emphasis on language as an active force, shaping the "mode of connection of ideas," pertained to human language, but it may be necessary to extend this statement to those languages we agree on and make use of by consent and convention. The more these new languages shape our world, the more we may be inclined to rank them among the defining metaphors of our time, bringing the analogy to language to flourish anew.

CONCLUSION

The chapters in this volume see the history of science through the lens of scientists' references to language. There is clearly an overlap between the history of our comprehension of language and the evolution of the ways scientists understand nature and pursue their work. Language has a history mirrored in the history of science, and natural philosophers and scientists have been selective in choosing the facets they regarded as important; thus language is wide open to various interpretations and uses.

Throughout the centuries enquirers into nature experimented with tongues in quite different ways and for quite different reasons. It is unlikely that we will stop doing so, not least because language is at the core of culture and encompasses the human as well as the natural sciences. In fact, despite various proposed alternatives—God as architect, engineer, clockmaker; the world as enigma, miracle, rational organism, logical system, or utilitarian mechanism —textual interpretations of the world, written either by DNA or by God, still prove powerful, especially with the rise of the biological paradigm.[37]

As these chapters show, various aspects of language can be taken as a measure of nature's activities. The justifications for drawing these analogies lay in the balance natural philosophers or scientists found between belief in the overall order of nature and therefore also of language, guaranteed either by God or—later—by evolution; and belief in manmade order and therefore in the perfection of language by convention. The more scientists drew analogies to conventional language, the more they were confronted with themselves: the way they conceived language and their own actions under the spell of their chosen languages. Here language was and still is the mirror of society and culture and therefore also of the natural sciences. In some moments transparent, in others opaque, and sometimes scientifically problematic, language may strike back when we let it become inaudible. At some stages of scientific work we would like to keep language at a distance; at others we would like to be close to it. In the end language is omnipresent.[38] This, however, does not mean—and here all the contributors agree—that nature works necessarily as language does.

CHAPTER TWO

The Linguistic Creation of Man: Charles Darwin, August Schleicher, Ernst Haeckel, and the Missing Link in Nineteenth-Century Evolutionary Theory

Robert J. Richards

While reflecting on various aspects of his new theory of species transformation, Charles Darwin (1809–82) conjured up a singing ape and then one groaning its desires while eyeing a well-proportioned member of the opposite sex. Such utterances, he mused, might have been the phonetic resources for primitive speech. The problem of language had captured Darwin's attention from a quite early period in his theorizing about species descent. His initial concern was to show that language—that most human of traits—had a natural origin and that it developed in genealogical and progressive fashion.[1] In a collection of notes, which he jotted down in 1837 shortly after returning from the *Beagle* voyage, he reflected on these putative features of language. On the very first page of this collection, he wrote, "all speculations on the origins of language.—must presume it originates slowly—if these speculations are utterly valueless—then argument fails—if they have, then language was progressive.—We cannot doubt that language is an altering element, we see words invented—we see their origin in names of People—Sounds of words—argument of original formation.—declensions &c often show traces of origin."[2]

A bit later he thought of that harmonious ape, when he queried himself: "Did our language commence with singing?" Were we originally like howling monkeys or chirping frogs? Alternatively, perhaps words arose out of expressions of emotion at certain events (for example, the ape with the opposite sex on its mind), or maybe from efforts at imitation of natural sounds.[3] These latter were the kinds of conjectures that Friedrich Max Müller (1823–1900), the great Oxford linguist, would later derisively call the "pooh-pooh" and "bow-wow" theories of language formation. Darwin worried, even at this early juncture, that if his views about language origins could not be sustained, then his whole argument regarding evolution might fail, since that argument could not then explain one of man's essential traits.

For the evolutionary thesis, no other trail lay open than the one Darwin initially began to follow. In the late 1860s, while focusing more determinately on constructing a theory of language, he came to rely in particular on his cousin Hensleigh Wedgwood (1803–91), who had endorsed a quasi-naturalistic account of linguistic development in his *On the Origin of Language* (1866); and while working on the *Descent of Man and Selection in Relation to Sex* (1871), Darwin made frequent inquiries of his cousin about the subject. Wedgwood had allowed that it was part of God's plan to have man instructed, as it were, by the natural development of speech. He argued that language began from an instinct for imitation of sounds of animals and natural events, which under "pressure of social wants" developed into a system of signs. According to Wedgwood, onomatopoeia served as the "vera causa" for a natural evolution of language.[4] Darwin embraced this confirmation of his original ideas, dispensing, of course, with the theological interpretation. In the *Descent of Man*, he mustered this naturalistic account of language acquisition to a surprising purpose.

The principal concern of the *Descent of Man*, as the title signals, is the evolution of the human animal, with all its distinctive properties, especially that of high intellect.[5] Darwin admitted that, as his friend Alfred Russel Wallace (1823–1913) had argued in the late 1860s, for survival, man's ape-like ancestors needed a brain hardly larger than that of an orangutan—actually not much larger, Wallace thought, than that exhibited by typical members of a Victorian gentleman's club. Wallace was reinforced in this conclusion by his turn toward spiritualism. He came to believe that man's ascent from the animal state occurred through the ministrations of higher, spiritual powers—a proposal that drove Darwin crazy.[6] Yet Darwin recognized the force of Wallace's objection. If a large brain, with all that such entailed, were not required for survival, then natural selection could not account for it. Darwin thus needed another way to explain the refinement and perfection of human intelligence. Language provided the instrument, although not in the way one might acknowledge today. In the *Descent of Man*, he argued in this fashion:

> The mental powers in some early progenitor of man must have been more
> highly developed than in any existing ape, before even the most imperfect

form of speech could have come into use; but we may confidently believe that the continued use and advancement of this power would have reacted on the mind by enabling and encouraging it to carry on long trains of thought. A long and complex train of thought can no more be carried on without the aid of words, whether spoken or silent, than a long calculation without the use of figures or algebra.[7]

Darwin proposed that man's apelike ancestors must have developed considerable intellectual capacity prior to breaking into the human range of intelligence. That animals displayed conspicuous understanding, approaching that of the human, no English huntsman seriously doubted. Even the great British idealist F. H. Bradley (1846–1924) remarked to a friend, "I never could see any difference at bottom between my dogs & me, though some of our ways were certainly a little different."[8] (This may say more about late-nineteenth-century British philosophy than about the abilities of English canines.) What was needed, in Darwin's view, to steam our animal ancestors across the Rubicon of mind was the engine of language. As language evolved through a natural development out of emotional and imitative cries, it would rebound on brain, promoting, as Darwin indicated, a more complex train of thought. Darwin would differ from contemporary neo-Darwinians, however. He believed that the complex patterns of thought that language stimulated would progressively alter brain structures and that these new acquisitions would produce an "inherited effect."[9] Language created human brain and, consequently, human mind. Darwin thus dissolved Wallace's objection to a naturalistic explanation of man.

From the beginning of his career to the end, Darwin believed in the inheritance of acquired characteristics. From our current perspective, we can see that he need not have argued in this fashion. He could have employed his own device of natural selection to explain the reciprocal pressures that mind and language might have exerted on one another to produce a continued evolution of both. Darwin did not appreciate that ever-more-complex language and thought might have had distinct survival advantages—for example, language might have served to weave together mutually supportive social networks for our protohuman ancestors. Like Wallace, he conceded that for sheer survival, our progenitors did not require a brain more advanced than that of, say, a great ape. Hence, in those cases in which natural selection seemed inapplicable, Darwin fell back on that device he always had at the ready—the inheritance of acquired characters.

Darwin's theory of the influence of language on developing mentality seems, at first blush, puzzling. This is not because of his employment of the idea of use-inheritance—common enough for his theory and his time. The puzzle rather arises because Darwin's proposal ran counter to the usual British empiricists' assumption that language merely expressed or mirrored ideas —it did not create them.[10] What then was the source of Darwin's conviction that language could mold human brain, could create human mind? In what follows I wish to argue that the ultimate source for his conception is to be found in German romanticism and idealism, especially in the work of Wilhelm von Humboldt (1767–1835), linguist and pedagogical architect of the University of Berlin, and of Georg Friedrich Hegel (1770–1831), Germany's greatest philosopher in the first part of the nineteenth century. German romanticism and idealism thus forged, I believe, a missing link in nineteenth-century evolutionary theory.

DARWIN AND THE LINGUISTIC RUBICON

Although Darwin realized that he would have to give an account of human mind and language if his general theory were to win the day, he kept all overt discussion of human evolution out of the book that first detailed his theory, the *Origin of Species* (1859). He simply forecast in the concluding chapter that "light will be thrown on the origin of man and his history."[11] The *Origin* is, nonetheless, larded with oblique but succulent references to human activity and history.[12] The case of language stands out among these. In his chapter on classification and systematics, for instance, Darwin observed, "If we possessed a perfect pedigree of mankind, a genealogical arrangement of the races of man would afford the best classification of the various languages now spoken throughout the world; and if all extinct languages, and all intermediate and slowly changing dialects, had to be included, such an arrangement would, I think, be the only possible one."[13] In this passage, Darwin recognized an isomorphism between language descent and human biological descent. Not only could the human pedigree serve as a model for tracing linguistic development, as he here emphasized, but also the reverse, as he implied, could be the case: the descent of language might serve as a model for the descent of man.

Darwin's suggestion about a similar genealogy for human beings and language passed casually through only one paragraph of the *Origin*. He himself did not really employ the model in any systematic way. His illustration of species evolution, the only graphic illustration in the *Origin*, was certainly not modeled on language development. The bare suggestion of this apparent isomorphism between the development of language and the development of human varieties, however, caught fire almost immediately. Although initially Darwin warmed himself contentedly in the blaze, his friend Charles Lyell (1797–1875) pushed him a little too close.

Lyell was a scientist out of whose brain, Darwin said, came half his ideas. Lyell immediately took up Darwin's suggestion about language descent and further advanced it in his book *The Antiquity of Man* (1863). Lyell had observed that although there were wide gaps between dead and living languages, with no transitional dialects preserved, competent linguists did not doubt the descent of modern languages from ancient ones. Therefore, gaps in the fossil record of species ought prove no more of an obstacle to transmutation theory than gaps in the record of language proved in linguistic theory. Moreover, the two kinds of descent should have a common explanatory account, he believed. So the formation and proliferation of languages were due, to quote Lyell, to "fixed laws in action, by which, in the general struggle for existence, some terms and dialects gain the victory over others."[14] He thus maintained that the processes of biological evolution could be likened to those of linguistic evolution—in both the more fit types were selected. Lyell, one of Britain's leading scientists of the time, in this way offered significant support for his friend's theory.

Lyell, however, could not cross the Rubicon. He thought the principle of natural selection was unable to account completely for the intricately designed fabric of language, even that of the more primitive languages of native groups. He judged—as Darwin groaned his great frustration—that natural selection of both language and life-forms could only be a secondary cause, operating under the guidance of higher powers. "If we confound 'Variation' or 'Natural Selection' with such creational laws," he cautioned, "we deify secondary causes or immeasurably exaggerate their influence."[15] Such a repair to higher wisdom, of course, eviscerated Darwinian nature of the fecund force with which the *Origin* invested it. And nature in Darwin's theory resonated of that romantic power of creative action and evaluation that it soaked up from German sources, especially from Alexander von Humboldt (1769–1859), whom Darwin incessantly read while on the *Beagle* voyage some years before.[16] But another German writer came to Darwin's attention in the mid

1860s, one whose analyses of language he found considerably more congenial than Lyell's and one whose ideas he would weave into his own theory of human evolution. This was August Schleicher (1821–68).

SCHLEICHER AND THE ROMANTIC THEORY OF LANGUAGE

Schleicher's Response to Darwin

Schleicher was a distinguished linguist working at the university in Jena. He had been urged by his good friend Ernst Haeckel (1839–1919) to read the German edition of the *Origin*.[17] Haeckel, who had recently converted to Darwinism, recommended the book because of Schleicher's horticultural interests.[18] But it was Schleicher the linguist who resonated more deeply to Darwin's work. He responded to Haeckel in an open letter, which he published as a small tract with the title *Die Darwinsche Theorie und die Sprachwissenschaft* (Darwinian theory and the science of language, 1863).[19] The book excited considerable controversy, evoking critically negative responses from the likes of Friedrich Max Müller and William Dwight Whitney (1827–94), and supportive efforts from Frederick Farrar (1831–1903).[20] In the *Descent of Man*, Darwin referred to his cousin Hensleigh Wedgwood and Farrar as sources for his ideas about evolutionary descent of language. He silently prescinded, as one might expect, from the fact that each of his sources reserved a role for the Creator. And he credited Schleicher as well. It was on Schleicher's thoroughgoing linguistic naturalism on which he principally depended for his theory of the constructive effect of language on mind.[21]

Schleicher indicated that contemporary languages had gone through a process in which simpler *Ursprachen* had given rise to descendent languages that obeyed natural laws of development. He argued that Darwin's theory was thus perfectly applicable to languages and, indeed, that evolutionary theory itself was confirmed by the facts of language descent. This last point was crucial for Schleicher, since it suggested the singular contribution that the science of language could make to the establishment of Darwin's theory. In the German translation of the *Origin*, Heinrich Bronn, the translator, had added an epilogue in which he allowed that Darwin's theory showed that descent was *possible* but that the Englishman had not shown that it was *actual*. According to Bronn, Darwin had provided no direct empirical evidence, only analogical possibilities.[22] Schleicher, like many other Germans, accepted Bronn's evaluation. He yet insisted that language descent, unlike the imaginative scenarios Darwin offered, could be proven—it was already an empirically established phenomenon. Moreover, the linguist's descent trees (*Stammbäume*) might be used as models for construing the evolution of plant and animal species.

Schleicher was quick to point out that the only graphic representation of descent in Darwin's *Origin* consisted of a highly abstract scheme, in which no real species were mentioned, only letter substitutes (see Figure 2.1). He contrasted this with a descent tree of the Indo-Germanic languages—his own graphic innovation—that he attached as an appendix to his tract (see Figure 2.2). Darwin had thus only represented a possible pattern of descent, while the linguist could provide a real pattern, empirically derived. Here, Schleicher believed, was a genuine contribution of linguistics to biological theory, a contribution that undercut Bronn's objection.

Schleicher maintained there were some four other areas in which the linguistic model could advance the Darwinian proposal. First, the linguistic system might display a "natural history of the genus Homo," because "the developmental history of languages is a main feature of the development of human beings." Second, "languages are natural organisms [*Naturorganismen*]" but have the advantage over other natural organisms since the evidence for earlier forms of language and transitional forms has survived in written records—there are considerably more linguistic fossils than geological fossils. Third, the same processes of competition of languages, the extinction of forms, and the development of more complex languages out of simpler roots all suggest mutual confirmation of the basic processes governing such historical entities as species and languages. Finally, since the various language groups descended from "cellular languages," language provides analogous evidence that more advanced species descended from simpler forms.[23]

Schleicher intended that these four complementary contributions of linguistics to biological theory should buttress an underlying conviction that received only vague expression in his *Darwinsche Theorie*, namely, that the pattern of language descent perfectly reflected human descent. The implicit justification for this proposition was simply that these two processes of descent were virtually the same. And this justification itself was grounded in

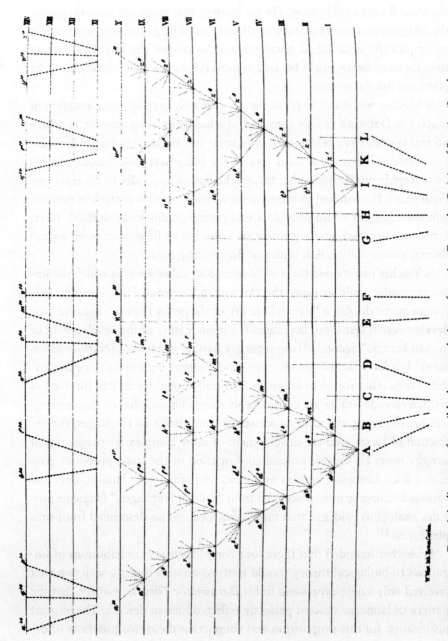

FIGURE 2.1 Darwin's diagram of possible descent relations of species. From the *Origin of Species* (London: John Murray, 1859).

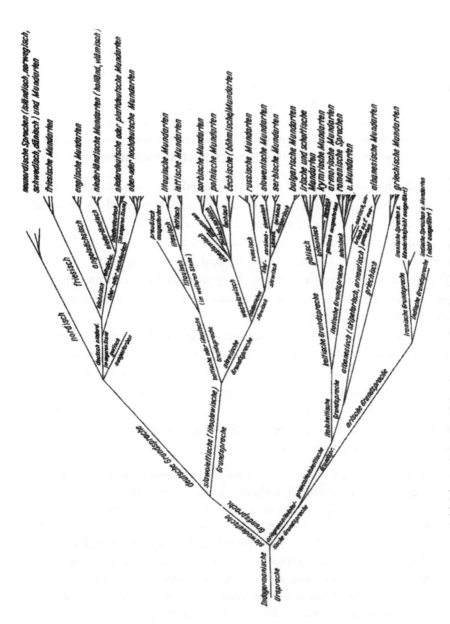

FIGURE 2.2 Schleicher's diagram of the descent relations of the Indo-Germanic languages. From his *Darwinsche Theorie und die Sprachwissenschaft* (Weimar: Böhlau, 1863).

the doctrine of monism that Schleicher advanced in his tract. The doctrine, as he formulated it, recognized:

> Thought in the contemporary period runs unmistakably in the direction of monism. The dualism, which one conceives as the opposition of mind and nature, content and form, being and appearance, or however one wishes to indicate it—this dualism is for the natural scientific perspective of our day a completely unacceptable position. For the natural scientific perspective there is no matter without mind [*Geist*] (that is, without that necessary power determining matter), nor any mind without matter. Rather there is neither mind nor matter in the usual sense. There is only one thing that is both simultaneously.[24]

For Schleicher, the doctrine of monism provided a metaphysical ground for his theory that the organism of language simply represented the material side of mind, which meant, therefore, that the evolution of one carried the evolution of the other. This organic naturalism had its roots in the German romantic movement. That movement rejected the mechanistic interpretation of nature and advanced the concept of *organism* as the fundamental principle in terms of which human mentality and all natural phenomena were ultimately to be understood.[25]

In a small work published two years after *Darwinsche Theorie*, Schleicher developed some further features of his complementary theories of linguistic and human evolution. In *Über die Bedeutung der Sprache für die Naturgeschichte des Menschen* (On the significance of language for the natural history of mankind, 1865), he argued that the superficial differences among human beings, which morphologists often exaggerated, proved simply insufficient to classify them. He observed:

> How inconstant are the formation of the skull and other so-called racial differences. Language, by contrast, is always a constant trait. A German can indeed display hair and prognathous jaw to match those of the most distinctive Negro head, but he will never speak a Negro language with native facility. . . . Animals can be ordered according to their morphological character. For man, however, the external form has, to a certain extent, been superseded; as an indicator of his true being, external form is more or less insignificant. To classify human beings we require, I believe, a higher criterion, one which is an exclusive property of man. This we find, as I have mentioned, in language.[26]

Since some languages were more perfect than others, this would provide a progressive arrangement of human varieties. Schleicher held, perhaps not surprisingly, that the Indo-Germanic and Semitic language groups were the most advanced, since they had features, such as tenses, declensions, and true noun and verb forms lacking in languages like the Chinese. By implication, he thus suggested that the most evolved human groups in the evolutionary hierarchy were those whose native languages were of the Indo-Germanic and Semitic families. Schleicher's justification for using language to classify human groups was quite simple: "The formation of language is for us comparable to the evolution of the brain and the organs of speech."[27] This was the position that Darwin endorsed, and it became for him a central feature of his evolutionary conception of mankind.[28]

Schleicher claimed that he himself had been convinced of the natural descent and competition of languages before he had read the *Origin of Species*. Although it is difficult to corroborate his assertion that he had previously urged a "Kampf ums Dasein" to explain language change, there is little doubt that he had affirmed language competition and descent as natural phenomena prior to reading Darwin and that he had used these concepts to argue for human evolution. Schleicher's argument, however, displays quite fascinating archaeological layers of earlier ideas.

Origin of Schleicher's Evolutionary Theory of Language and Mind

Schleicher was born February 19, 1821, in Meiningen (southwest of Weimar in the Thuringian Forest) to a physician with a taste for nature and his musically talented wife.[29] The professors of his gymnasium cultivated exotic languages but did not, amazingly, have high hopes for this particular pupil. In fall 1840, Schleicher began the curriculum in theology at Leipzig and the next semester traveled to Tübingen for more of the same. At Tübingen his passion for the transcendent found secular liberation in Hegel's writings, which had been recently collected by his students (1832–40), with many works appearing for the first time. Schleicher also began acquiring languages at a frightening rate: Arabic, Hebrew, Sanskrit, and Persian initially. With the reluctant permission of his father, he went to Bonn, in 1843, to devote himself to the study of classical languages. There he entered the seminar conducted by the famous classical philologists Friedrich Ritschl (1806–76) and Friedrich Welcker (1784–1868), who introduced him to the linguistic ideas of Wilhelm von Humboldt.[30] Although of oscillating health while at Bonn, Schleicher yet braced his study with participation in gymnastic competitions, a recreation that he and Haeckel would later together pursue with avidity. He received a doctorate in 1846 and would normally have then spent time as a professor in a gymnasium before pursing further study. He fell, however, under the protective wing of Prince Georg von Meiningen, who, admiring of his landsman's talents, arranged for a generous stipend. The money enabled Schleicher to continue his study during a period of two years of extensive travel (1848–50).

In the summer of 1848, after the February Revolution and the establishment of the Second Republic, Schleicher journeyed to Paris to continue his linguistic research in the Bibliothéque Nationale. He augmented his income during this sojourn by serving as correspondent to the *Allgemeine Zeitung* (Augsburg) and the *Kölnische Zeitung*. He reported on the fluctuating political events occurring in Paris and a bit later in Vienna, as revolution spread to the capital of the Hapsburg Empire. Schleicher's reports, tinged with the sympathetic color of a liberal democrat, followed the fate and abortive efforts to establish a republic in the Germanies.[31] In addition to his political reporting, Schleicher managed to produce a number of important linguistic studies, which elicited a call from the University of Prague to the position of extraordinary professor. Three years later, he advanced to ordinary professor of German, comparative linguistics, and Sanskrit. He remained in Prague until 1857, when he received an offer to return to his own land. He accepted a position in the philosophy faculty at Jena, the venerable university that two generations earlier, at the turn of the century, had nurtured the romantic movement, serving as redoubt for the likes of Schiller, Fichte, the brothers Schlegel, Schelling, Hegel, and with Goethe right down the road at Weimar. Jena was also the university of Schleicher's father, Johann Gottlieb (1793–1864), who in the summer of 1815 helped found the first Burschenschaft, the student organization that agitated for democratic reform and political unity.[32] In the 1850s, the university looked back to a glorious past and forward to a financially precarious future.

Although he initially had high hopes for his time in Jena, undoubtedly recalling his father's stories of revolutionary days at the university, Schleicher quickly came to feel isolated from his colleagues, whose conservative considerations bent them away from the more daring of his own approaches both in linguistics and politics. The poor finances of the university, making scarce the necessities of scholarship, did not improve his attitude. A friend remembered Schleicher remarking that "Jena is a great swamp and I'm a frog in it."[33] The frog was saved from wallowing alone in his pond when Ernst Haeckel arrived at the university in 1861. They took to one another immediately and remained fast friends through the rest of Schleicher's short life. He died in 1868, at age forty-eight, apparently of a recurrence of tuberculosis.

In 1848, after he returned to Bonn from research in the revolution-torn Paris, Schleicher saw published his first monograph, *Zur vergleichenden Sprachengeschichte* (Toward a comparative history of languages).[34] This work framed the theory that would guide him through the rest of his career. In it, he distinguished three large language families by reason of their forms: isolating languages, agglutinating languages, and flexional languages. Isolating languages (for example, Chinese and African) have very simple forms, in which grammatical relationships are not expressed in the word; rather, the word consists merely of the one-syllable root (with position or pitch indicating grammatical function). Because of their simple structure, these languages cannot, according to Schleicher, give full expression to the possibilities of thought. Agglutinating languages (for example, Turkish, Finnish, Magyar) have their relational elements tacked on to the root in a loose fashion (indeed, the relational elements themselves are derived from roots). Flexional languages (for example, the Indo-Germanic and Semitic families) are the most developed. Roots and relations form an "organic unity," according to Schleicher.[35] So, for example, the Latin word "scriptus" has "scrib" as the root or meaning; "tu" expresses the participial relationship; and "s" indicates the nominative relationship. Schleicher believed that even the most highly developed languages, the flexional group, originated from a simpler stem, much like the Chinese, but continued to develop into varieties with more perfect forms. Isolating and agglutinating languages, by contrast, simply did not have the potential to move much beyond their more primitive structures.

Schleicher regarded these three language forms as exhibiting an internal, organic unity. Indeed, he compared them to natural organisms of increasing complexity: crystals, plants, and animals, respectively.[36] Such comparisons had the authority of those linguists upon whom Schleicher most relied: Wilhelm von Humboldt, Franz Bopp (1791–1867), and August Wilhelm Schlegel (1767–1845). These researchers, all tinged by the romantic movement, employed the organic metaphor with alacrity.[37] Schleicher, however, did suggest an important disanalogy between languages and biological organisms. Languages had a developmental history, whereas biological organisms, although they came to exist through a gradual process, did not alter once they were established. They essentially had no history. At least this was Schleicher's view in 1848.

In 1850, Schleicher completed a large monograph systematically describing the languages of Europe, his *Die Sprachen Europas in systematischer Übersicht* (The languages of Europe in systematic perspective). He now explicitly represented languages as perfectly natural organisms that could most conveniently be described using terms drawn from biology—for example, genus, species, and variety.[38] Some of his contemporaries, as well as later linguists, have thought Schleicher's conception of language as a natural, law-governed phenomenon to be erroneous, a denial of man's special status. Such critics then (and now) failed to understand that this was not a denigration of the *geistlich* character of language; rather, it was, in the romantic purview, an elevation of the natural.[39] Romantics and idealists—such as Schelling, Schlegel, and Hegel—deemed nature simply the projection of mind. Schleicher, then, did not reduce in vulgar fashion the spiritual dimension of language to some nonanimate concourse of atoms in the void.

In his *Die Sprachen Europas*, Schleicher suggested (but did not yet graphically illustrate) that the developmental history of the European languages could best be portrayed in a *Stammbaum*, a stem-tree or developmental tree. He first introduced a graphic representation of a *Stammbaum* in articles published in 1853, representations that indeed looked like trees (see Figure 2.3).[40] By the time of the publication of his *Deutsche Sprache*, seven years later (1860), he had begun to use *Stammbäume* rather frequently to illustrate language descent (see Figure 2.4). Schleicher is commonly recognized as the first linguist to portray language development using the figure of a tree.[41] Certainly he thought carefully about how illustrations could make more clear, more intuitive the descent relations that purportedly obtained among languages. So, for instance, he used the angular distance separating the branching of the *Stammbaum* to suggest the morphological distances of daughter languages (see Figure 2.5).[42] Such illustrations, so intuitively seductive, acted as tacit arguments for the theory they depicted.

In *Deutsche Sprache*, Schleicher reiterated the argument of *Die Sprachen Europas* that more recent languages had descended from *Ursprachen* and that their descent conformed to natural laws. He now, however, started to formulate those laws (for example, "When two or more branches of a language stem [*Sprachstamm*] are quite similar, we may naturally conclude that they have not been separated from each other for very long").[43] He also

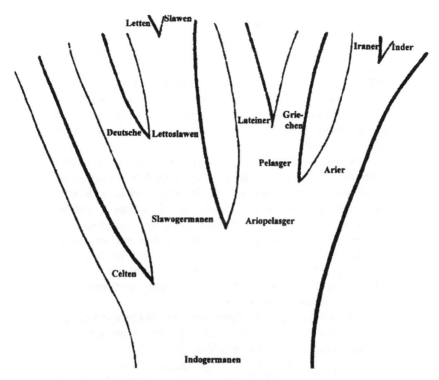

FIGURE 2.3   Schleicher's first diagram of language descent. From "Die ersten Spaltungen des indogermanischen Urvolkes," *Allgemeine Zeitschrift für Wissenschaft und Literatur* (August 1853).

made explicit a vague notion that had been floating around in his earlier works. He argued that the descent of languages paralleled the descent of man, that indeed, more primitive animal forms achieved their humanity precisely in acquiring language. As he expressed it: "According to every analogy, man has arisen out of the lower forms, and man, in the proper sense of the word, first became that being when he evolved [*entwickelte*] to the point of language formation."[44] Schleicher further maintained that since human languages were polygenic in origin, so was man. That is, he believed that there was no one *Ursprache* whence the other languages descended; rather, there were many *Ursprachen*, each having developed in different geographical regions out of cries of emotion, imitation, and ejaculation. Since language and thought were two sides of the same process, as language groups developed and evolved independently of one another, so did the different groups of human beings who spoke them.[45]

indogermanischen Ursprache ausschied, also am längsten ein selbst=
ständiges Leben führte und so sich individueller entwickelte. Sla=
wisch und Litauisch stehen sich aber außerordentlich nahe, sie sind
erst sehr spät aus einer gemeinsamen Grundsprache, der slawo=
lettischen hervorgegangen. Die slawodeutsche Grundsprache schied
sich also zuerst in deutsch und slawolettisch, dieses sodann in let=
tisch und slawisch.

So sind wir denn durch genauere Betrachtung der Verwandt=
schaftsverhältnisse der einzelnen indogermanischen Grundsprachen
(Familien) und durch die auf die Grundlage solcher Erkenntniß
nothwendig sich aufbauenden Schlüsse auf die ältesten Sprachthei=
lungen zu einer genaueren Einsicht in unsere sprachliche Vor=
geschichte gelangt; nehmen wir noch hinzu, daß wir mit gleicher
Sicherheit die indogermanische Ursprache selbst noch in ihrem Wer=
den zu begreifen im Stande sind, so wird man den Leistungen
unserer noch so jungen Disciplin Anerkennung, ja Bewunderung
wohl kaum versagen können.

Die über das successive Hervorgehen der acht indogermanischen
Grundsprachen aus der gemeinsamen Ursprache gewonnenen Ergeb=
nisse mag folgendes Schema veranschaulichen.

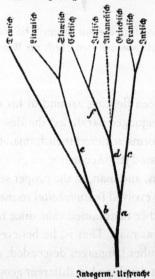

In diesem Schema bedeutet u
die asiatisch = südeuropäische Grund=
sprache, b die nordeuropäische (slawo=
deutsche) Grundsprache, Sprachen, die
beide durch die erste Theilung der
indogermanischen Ursprache entstun=
den; c ist die asiatische (arische)
Grundsprache, d die südeuropäische
(pelasgoceltische, gräcoitaloceltische)
Grundsprache, c und d sind also die
beiden Töchter von a, in welche es
sich auflöste; das Albanesische wag=
ten wir als frühe Abzweigung vom
griechischen Aste kaum anzudeuten;
f ist die italoceltische Grundsprache,
das übrige ist durch die beigesetzten
Namen an der Zeichnung selbst an=
gegeben.

Indogerm. Ursprache

Schleicher, deutsche Sprache.   6

FIGURE 2.4   Schleicher's *Stammbaum* of the Indo-Germanic languages. From
Die Deutsche Sprache (Stuttgart: Cotta'scher Verlag, 1860).

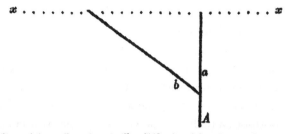

Die Grundſprache A theilt ſich in die Sprachen a und b in
der beſchriebenen Weiſe nämlich ſo, daß der Theil des Sprach=
gebietes b ſtärkeren Veränderungen unterliegt als der mit a bezeich=
nete. Bis zum Durchſchnitt xx hat alſo b ſich viel weiter von
A entfernt als a, und dieß macht eben unſer Schema dadurch an=
ſchaulich, daß es bx ſtärker von der geraben Richtung abweichen
läßt als ax, das mehr als eine directe Fortſetzung von A er=
ſcheint.

FIGURE 2.5   Schleicher's graphic method of intuitively showing the greater
divergence of daughter language *b* from the mother language *A* and the more
lineally descended daughter language *a*. From his *Die Deutsche Sprache* (Stuttgart:
Cotta'scher Verlag, 1860).

Schleicher on the Evolution of Man, the Language User

Prior to having read Darwin, Schleicher seems to have already convinced
himself that human beings had derived from lower animals. Certainly from
the beginning of the nineteenth century, several German biologists—for example, Gottfried Treviranus (1776–1837), Friedrich Tiedemann (1781–1861),
and Johann Meckel (1781–1833), stimulated by Jean-Baptiste de Lamarck
(1774–1829)—had become full-blown evolutionists.[46] But was Schleicher full-
blown before 1859? His argument for human descent depended on the identification of language and thought. The linkage itself has a venerable history.
Authors as far back as Plato understood language and thought to have a close
relationship. Johann Gottfried Herder (1744–1803), an author every German
intellectual of the first half of the nineteenth century assiduously read, con-
tended, in a prizewinning treatise of 1772, that language was necessary for
thought, "that indeed the first and most elementary application of reason
cannot occur without language." Contrary to the creationists, Herder urged
that speech arose gradually in human groups, initially through imitation of

natural sounds. "No Mercury and Apollo," he protested, "descend from the clouds as by opera machinery—the whole, many-sounding, divine nature is the language teacher and Muse for man."[47] Schleicher would endorse the notion that languages first arose out of imitation of natural sounds, but he conceived an even tighter relationship between language and thought, namely, that of virtual identity.[48] In so doing, he seems proximately to have developed a theoretical position initially laid down by Wilhelm von Humboldt in his *Über die Kawi-Sprache auf der Insel Java* (On the Kawi-language on the island of Java, 1836).

In his introduction to the *Kawi-Sprache*—a work often cited by Schleicher —Humboldt argued for the intimate relation between thought and language. He formulated the relationship in this way: "Just as without language no concept is possible, so likewise without language there is no object for the soul, since it is only by means of the concept that any external object can express its complete essence for the soul."[49] Humboldt also suggested, equally darkly, that the descent (*Abstammung*) of language "joined in true and authentic union with physical descent."[50] It would take only slightly more conceptual boldness for Schleicher to conclude, as he forthrightly did, that the descent of language paralleled the descent of thought or mind. Thus the conclusion of *Deutsche Sprache*: with the evolution of different languages comes the evolution of different kinds of human beings.

Yet one can still ask: Did Schleicher's conclusion amount to endorsing something like the Darwinian thesis before Darwin? A clue to the answer to this question can be gleaned from examining a most curious theory in *Deutsche Sprache* concerning the evolution of language in human groups.

Schleicher argued that human beings, in their acquisition of language, went through three periods of development: a prelinguistic period, a prehistorical period of language emergence and development, and then a historical period of language decline. In the earliest stage, when no true languages existed, neither did human beings—since without language there could be no human thought. In the next, the prehistorical phase of earth's history, languages (and thus human beings) began to develop. During this period, many different language groups sprang into existence and many died out— indeed, most languages went extinct before achieving their full potential. Others, however, began to spread from one region to another. When languages achieved their maturity, human beings entered the historical period, during which they became self-conscious through the medium of historical understanding. However, with the advent of the historical period, no fundamentally new languages arose. Indeed, during this time, languages began to decline, to devolve! Words started to fall out, forms became simplified, and grammatical relations were lost. Thus Greek and Latin have a much richer store of grammatical forms than modern languages descended from them. Yet, during this historical period, culture and reason dramatically advanced. Schleicher's scheme of language evolution, with its initial progress and then devolutionary decline, seems perfectly paradoxical—that is, until its roots are uncovered.

The fundamental features of this scheme appeared in Schleicher's first monograph, where it is obvious that the basic conception came from Hegel. In the *Zur vergleichenden Sprachengeschichte*, Schleicher depicted the three language forms (the isolating, agglutinating, and flexional) as moments in the development of the "World Spirit" (*Weltgeist*). The Spirit, in the Hegelian view, strove to realize itself, to become fully self-conscious. This striving would be instantiated in the development of human mentality and revealed in language formation. Thus languages would move through dialectical stages, from simple expressions of meaning (in isolating languages), to the structural antithesis in languages that loosely joined meaning and relationships (agglutinating), to a higher synthesis in the "organic unity" of the word, characteristic of the flexional groups—the Semitic and Indo-Germanic. "Whatever we recognize as significant in any sphere of the human spirit," Schleicher averred, "has blossomed from one of these two groups [that is, Semitic and Indo-Germanic]."[51] In Hegel's view, one explicitly adopted by Schleicher, during the prehistorical period the World Spirit established the intellectual resources —namely, highly developed languages—so as to begin the process of historical self-reflection and the attainment of freedom. Once the process had begun, however, the energies required for the refined articulation of language began to be employed in the development of rational laws, state governments, and the aesthetic products of advanced civilization. "Hegel thus recognized," according to Schleicher, "the fact that the formation of languages and history cannot take place at the same time, that in the advance of history, rather, language must be worn down."[52]

In Hegel's *Vorlesungen über die Philosophie der Geschichte* (Lectures on the philosophy of history, 1840), from which Schleicher initially drew his theory, the prelinguistic period of human existence is represented as nonetheless *potentially human*, with the "germ or drive" to reflective consciousness already built in.[53] Hegel certainly stopped short of a full-blown biological evolutionism, and this may be where Schleicher himself stopped in *Deutsche Sprache*. Yet, there can be little doubt that Schleicher was brought to the conceptual brink of biological transformation theory by Humboldt and Hegel— even if, after 1848, Hegel's name never again appeared in Schleicher's texts. The reading of Darwin's *Origin of Species*, under Haeckel's tutelage, provided the shove for one who was ready to take the plunge into a new conceptual sphere.

Schleicher's own evolutionism obviously went through stages of development, finally resting in his adoption of Darwinism in language and human evolution. One significant index of Darwin's impact on Schleicher's linguistic ideas was the absence of the theory of language decline in his *Darwinsche Theorie*. Darwin's theory of development was thoroughly progressivistic; therefore, it would have been anomalous to suggest that the natural selection of languages led to a devolution of language. Yet Schleicher would have realized that his original assumption of the perfection of ancient languages was one still widely shared by linguists and cultural critics in love with the classics. He would appear to have only one recourse, which he took—namely, silence. For the most part, however, Darwin's ideas simply overlaid the fundamental features of Schleicher's prior evolutionary project, which derived from the work of those individuals immersed in German romanticism and idealism—especially Humboldt and Hegel. They had initially argued that the model of organic growth formed the basic category for understanding the development of consciousness. Their fundamental metaphysical view was monistic—mind and matter expressed two features of an organic Urstoff— and this sort of monism became the assumption of evolutionists during the latter half of the nineteenth century, especially of Haeckel.

HAECKEL'S THEORY OF THE LINGUISTIC EVOLUTION OF MAN

Ernst Haeckel, to whom Schleicher's *Darwinische Theorie* had been addressed, had converted to Darwinism in 1860, virtually as soon as he read the German translation of the *Origin*. At the time, he was working on his habilitation, in which he would describe and systematically classify the radiolaria, simple one-celled creatures that inhabited the oceans and exuded an exoskeleton.[54] Darwin's theory helped him make sense of the myriad of families, genera, and species these creatures displayed. Haeckel, like Schleicher, had been ready for such a theory as Darwin's; he too was thoroughly imbued with romantic ideals. His letters to his fiancée—written while working on his habilitation in southern Italy—are smeared with quotations from Goethe. The romantic élan so took his soul in thrall that he contemplated giving up his scientific work for that of the life of a painter and free spirit. For a time he wandered over the island of Capri with a poet friend, who almost seduced him, quite literally, away from his eventual career as a university professor. It was only the thought of his fiancée, with whom he was deeply in love, and the realization that the life of a Bohemian did not pay very well that steeled him to finish his habilitation and return to Jena.[55]

Haeckel remained at Jena throughout his career and under his influence during the last half of the nineteenth century, the university became a bastion of Darwinian thought. Schleicher, who quickly slid to the Darwinian side under his friend's guidance, in turn contributed to Haeckel's own version of Darwinism, a version that became part of the standard view through the early years of the twentieth century. Schleicher made several significant contributions. First, he confirmed, from a quite different perspective, Darwin's theory, and thus supported Haeckel in what would become a comprehensive scientific philosophy. Second, he solidified for his friend that important metaphysical vision that became the basis for evolutionary theory in the latter half of the nineteenth century, namely monism.

Monism could support a variety of philosophical refinements. For instance, the American pragmatists William James (1842–1910) and John Dewey (1859–1952) both avowed monism. Henri Bergson (1859–1941) also claimed that metaphysical doctrine, as did most other evolutionists. Haeckel himself elevated the doctrine into a "monistic religion," as he termed it.[56] The philosophy of monism could be given, as the works of these individuals suggest, different spins, different emphases. Haeckel always reminded his readers that anything called *Geist* had a material side. So, for example, under the rubric of monism in his *Natürliche Schöpfungsgeschichte* (The natural history of creation, 1868), which was a popular version of his fundamental theoretical work, *Generelle Morphologie der Organismen* (The general morphology of organisms, 1866), Haeckel insisted that "the human soul has been gradually formed through a long and slow process of differentiation and perfection out of the vertebrate soul."[57] Or, as he also put it, "Between the most highly developed animal soul and the least developed human soul there is only a

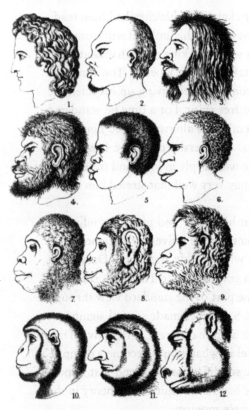

Natürliche

Schöpfungsgeschichte.

Gemeinverständliche wissenschaftliche Vorträge
über die

Entwickelungslehre

im Allgemeinen und diejenige von

Darwin, Goethe und Lamarck

im Besonderen, über die Anwendung derselben auf den

Ursprung des Menschen

und andere damit zusammenhängende

Grundfragen der Naturwissenschaft.

Von

Dr. Ernst Haeckel,

Professor an der Universität Jena.

Mit Tafeln, Holzschnitten, systematischen und genealogischen Tabellen.

Berlin, 1868.
Verlag von Georg Reimer.

Die Familiengruppe der Katarrhinen (siehe Seite 555).

FIGURE 2.6   Frontispiece and title page of Ernst Haeckel's *Natürliche Schöp-fungsgeschichte* (Berlin: Reimer, 1868).

quantitative, but no qualitative difference."[58] Indeed, Haeckel thought that the mental divide separating the lowest man (the Australian or Bushman) and the highest animal (ape, dog, or elephant) was smaller than that separating the lowest man from the highest man, a Newton, a Kant, or a Goethe.[59] Haeckel regarded differences among men as so significant, that he thought humankind should be classified not simply into different races or varieties of one species, but into some nine separate species of one genus (see Figure 2.7).

Morphological similarities led Haeckel to argue that human beings evolved through a kind of bottleneck, that of the narrow-nosed apes (see Figure 2.6). There must have been, according to Haeckel, an *Urmensch*, or *Affenmensch* —an ape man—that stemmed from the *Menschenaffen*—the menlike apes.

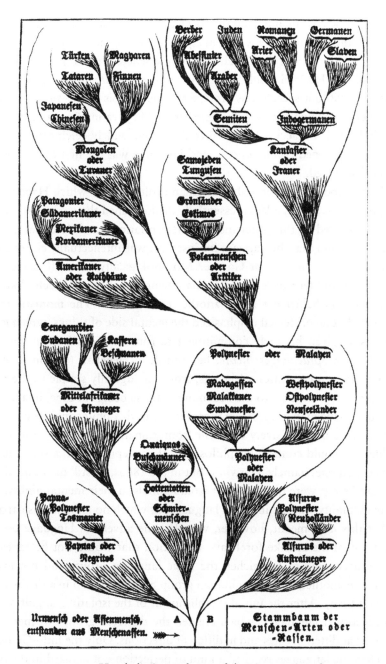

FIGURE 2.7 Haeckel's *Stammbaum* of the nine species of men.
From his *Natürliche Schöpfungsgeschichte* (Berlin: Reimer, 1868).

This was the missing link, and the currency of this idea is due to Haeckel. He thought the *Affenmensch* would likely have come either from Africa or perhaps from the area of the Dutch East Indies, where the orangutan was to be found. Later, Haeckel would name this Ur-ancestor *Pithecanthropus alalus*—ape-man without speech. His protégé Eugène Dubois (1858–1940), a Dutch army doctor, actually found *Pithecanthropus* in Java in 1891; and the missing link, which Haeckel had predicted, became widely celebrated.[60] It was later re-christened *Homo erectus*, and Java man was the first of his remains to be discovered.

The unspoken question about human evolution, for which Haeckel had a spoken answer, was: What essentially distinguished the various species of men, what led to this great mental differentiation—a differentiation that persuaded him that the Papuan, for instance, was intellectually closer to the apes than to a Newton or Goethe? Morphologically, after all, aside from skin color and hair differences, human beings were pretty much alike. On this question Schleicher made another contribution. The monistic metaphysics that he professed emphasized the mental side of things, which is not surprising given his early commitment to romantic idealism. In *Zur vergleichenden Sprachengeschichte*, he argued, in Hegelian fashion, that the systematic representation of beings, from the logically simple to the more complex, was identical to the becoming of those beings in time, in a kind of evolutionary emanation. Animal cognition, in this philosophical consideration, remained decisively different from human mentality. By the 1860s, Schleicher could confirm his philosophical conception with a scientifically articulated one, namely, Darwin's. But in the 1860s, he still maintained that human beings were quite distinct from animals in their mental ability. Human mentality was exhibited in language, of which no animal was capable. What this now meant, however, was that the advent of language created man out of his apelike forebears, a creation that would not be repeated. Since, according to Schleicher, the basic language groups did not evolve from one another, each protohuman group became human in a distinctively different way. After the initial establishment of the isolating, agglutinating, and flexional languages, which created the different groups of men, they evolved at different rates and in different directions. Only the Indo-Germanic and Semitic languages reached a kind of perfection not realized in the other groups. Here, then, was Haeckel's solution to the evolution of the various human species.

In the *Natürliche Schöpfungsgeschichte*, Haeckel maintained that human beings had a quasi-monogenic origin in *Pithecanthropus*. He imagined that these original protomen evolved on a continent that now lay sunken in the Indian Ocean, somewhere between Malay and South West Africa, and that these primitive *Urmenschen* eventually split into two groups, which migrated respectively toward both east and west. Later he would call this fanciful continent "Atlantis" or "Paradise," with the full irony of that latter name in mind. Although the human physical frame could be traced back to this one kind of ape-man, Haeckel yet maintained that in a proper sense, the human species were polygenic, as Schleicher had suggested:

> We must mention here one of the most important results of the comparative study of languages, which for the *Stammbaum* of the species of men is of the highest significance, namely that human languages probably had a multiple or polyphyletic origin. Human language as such probably developed only after the species of speechless *Urmenschen* or *Affenmenschen* had split into several species or kinds. With each of these human species, language developed on its own and independently of the others. At least this is the view of Schleicher, one of the foremost authorities on this subject. . . . If one views the origin of the branches of language as the special and principal act of becoming human, and the species of humankind as distinguished according to their language stem, then one can say that the different species of men arose independently of one another.[61]

The clear inference is that the languages with the most potential created the human species with the most potential. And, as Haeckel never tired of indicating, that species with the most potential—a potential realized—was that constituted by the Semitic and Indo-Germanic groups, with the Berber, Jewish, Greco-Roman, and Germanic varieties in the forefront.[62] Their vertical position on the human *Stammbaum*, indicated the degree of their evolutionary advance (see Figure 2.7 above).

But Schleicher's greatest and lasting contribution to evolutionary understanding may simply be his use of a *Stammbaum* to illustrate the descent of languages. Not long after Schleicher published his open letter, Haeckel finished his magnum opus, his synthesis of evolutionary theory and morphology, his large two-volume *Generelle Morphologie der Organismen*. The end of the second volume included eight tables of phylogenetic trees. Although there are some vague antecedents for the graphic use of treelike forms for the expression of descent relationships, Haeckel quite obviously took his

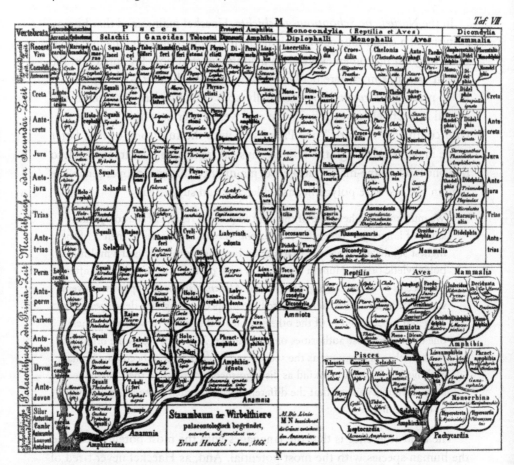

FIGURE 2.8   Haeckel's *Stammbaum* of the descent of vertebrates. From his *Generelle Morphologie der Organismen* (Berlin: Reimer, 1866).

inspiration from his good friend Schleicher.[63] And Haeckel's *Stammbäume* have become models for the representation of descent ever since.

Haeckel's tree of vertebrates (see Figure 2.8) might be compared with both Darwin's diagram and Schleicher's. Unlike Darwin's but like Schleicher's, Haeckel's illustration shows a single origin of the vertebrate phylum, although each of the major phyla (for example, Mollusca, Articulata, and so on), he maintained, had independent origins. And, of course, again unlike Darwin's but like Schleicher's, Haeckel's *Stammbaum* depicts actual species, the extinct and the extant. Schleicher's tree captured both time, marked as the distance from the Indo-Germanic *Ursprache*, and morphological differentiation, represented by the separation of the branches. Haeckel's diagram depicts this too. Haeckel's tree has an added feature, of course: it actually looks like a tree, whereas Darwin's and Schleicher's sketches are merely line drawings. This might seem, at bottom, a trivial difference, arising from the fact that Haeckel was an accomplished artist. Certainly his talent made the depiction possible. But the living, branching, gnarled, German oak functioned as a kind of graphic rhetoric: it vividly displayed the tree of life, in all its gothic and romantic textures. In the case of all three authors, but with increasing vivacity, a visual argument was made, which with Haeckel had become a powerful, if silent, linking of the very newest theory in biology with the traditions well established at Jena of German romanticism.

CONCLUSION

During the mid-1860s, Darwin's great friend Alfred Russel Wallace had undergone a conversion to spiritualism, on the basis of experimental evidence, to be sure.[64] In a review article in 1869, Wallace fortified his conviction with some powerful arguments about natural selection's insufficiency to account for man's big brain.[65] Sheer survival, he thought, simply did not require the intellectual capacity demonstrated by even primitive men. Darwin, in some horror, responded to his friend's article: "But I groan over Man—you write like a metamorphosed (in retrograde direction) naturalist, and you the author of the best paper that ever appeared in the *Anthropological Review!* Eheu! Eheu! Eheu!"[66] Darwin, nonetheless, saw the force of Wallace's argument and thus the vexing problem it posed: how to explain the complex mind and big brain of human beings. But during the mid-1860s, another kind of argument came to his attention, one that held the key to the evolutionary resolution of the problem. The argument was Schleicher's for the linguistic creation of man, and it came to Darwin's attention through several sources.

Darwin studied Schleicher's *Darwinsche Theorie*, which he then used and cited in his own formulation in the *Descent of Man*. He got two other doses of Schleicher's views more indirectly. Frederick Farrar—whom Darwin named along with his cousin Hensleigh Wedgwood and Schleicher as contributing to his conception of language—had made Schleicher's theories known to the British intellectual community through a comprehensive account in the journal *Nature*.[67] Schleicher's conceptions also got conveyed to Darwin through a gift of Haeckel's *Natürliche Schöpfungsgeschichte*, which the author sent in

1868. Darwin wrote to a friend after reading Haeckel's work that it was "one of the most remarkable books of our time."[68] Darwin's notes and underlining in the book are quite extensive. He was particularly interested, as shown by his scorings and marginalia, in Haeckel's account of Schleicher's thesis in *Über die Bedeutung der Sprache für die Naturgeschichte des Menschen*.[69] Here Darwin had a counterargument to Wallace's, one by which he could solidify an evolutionary naturalism: language might modify brain, increasing its size and complexity, with such acquisition becoming a permanent, hereditary legacy.[70] The irony, of course, is that Darwin's evolutionary naturalism obtained its support, via Schleicher, ultimately from Wilhelm von Humboldt and Georg Friedrich Hegel, two foremost representatives of German romanticism and idealism, the movements that forged the missing link in nineteenth-century evolutionary theory.

CHAPTER THREE

Is the Notion of Language Transferable to the Genes?

Stephanie Suhr

Dedicated to the memory of my father, Erich Meng

Coined in the Middle Ages, the metaphor of the *book of nature* or the *book of the world* expresses the human desire to reveal the secret of life and its underlying principles.[1] Today, DNA is interpreted as "text" to be read and understood in order to understand living nature: "What has been learned thus far . . . is analogous to having figured out the alphabet, the structure of some words, and a few rudiments of the grammar of a language that is yet to be deciphered. . . . Yet at some future time we will know enough to read the language fluently."[2]

Although the term *book of nature* is clearly a metaphor, things seem different when we talk about the "language of genes" or "molecular-genetic language." Bernd-Olaf Küppers, for example, claimed, "When we denote the principle of molecular information storage in the DNA molecules of an organism as a molecular-genetic language, there is more in this than mere metaphor."[3] This statement must be regarded critically.

The notion of *metaphor* remains under discussion and continues to defy "every encyclopedic entry" despite voluminous publications on the subject.[4] I will return to the original Greek word *metapherein* (to carry beyond), according to which a metaphor is the "'transferred' meaning of a word, which is no longer used in its 'actual' sense."[5] The metaphor appears as a picture, as the unreal use of a term. It might superficially appear that a metaphor relies on an arbitrary comparison, but this is simplistic. Eco, following Aristotle, pointed out the cognitive function of metaphors: "Creating metaphors 'is a sign of a natural disposition of the mind,' because knowing how to find good metaphors means perceiving or grasping the similarity of things between each other."[6] Between the "unreal metaphor" and an actual expression there is always a *tertium comparationis*, a similarity.

If it is Küppers's claim that the notion of a "molecular genetic language" is more than mere metaphor, this entails going beyond comparisons of a few selected elements of DNA and human language, beyond the merely heuristic use of the comparison, and demonstrating that direct equivalents between the structure and the functioning of both systems can be established.

The question is thus whether the genetic apparatus of the cell indeed has a linguistic character or one can only speak of it in a pictorial sense, in a transferred sense. Starting from a short presentation of the historical process that led to the conception of life as a genetic text, I will discuss some central works that argue for a deeper analogy between language and the genetic realm. A discussion of the notion of *language* is indispensable in this connection. I will also discuss the several investigations that in recent years have aimed to test the "language character" of DNA in practice using linguistics techniques. By looking more closely at the uses of metaphors and analogies, I will examine the possibility of attributing a linguistic character to DNA and the molecular-genetic realm.

THE ORIGINS OF THE METAPHOR OF LANGUAGE
IN INFORMATION THEORY

Points of departure for the adoption of linguistic and semiotic notions and metaphors have been, on the one hand, the physicist Erwin Schrödinger's book *What Is Life?* (1944), and, on the other, cybernetics, information theory, computer technology, and the control and communication systems developed after the Second World War.[7] Lily Kay stresses the much larger influence of the latter.[8] Norbert Wiener in his book *Cybernetics: Or Control and Communication in the Animal and the Machine* (1948) used the term "message" (by which he understood "a discrete or continuous sequence of measurable events distributed in time") also for living organisms in his book *The Human Use of Human Beings: Cybernetics and Society* (1950): "There is no fundamental absolute line between the types of transmission which we can use for sending a telegram from country to country and the types of transmission which at least are theoretically possible for transmitting a living organism such as a human being."[9] Here the notion of *message* is no longer used metaphorically; "Life is no longer 'like a message'—it *is* a message."[10] In reaction to Wiener's thesis geneticists and molecular biologists began in the early 1950s to redefine the organism as a cybernetic system and to phrase their results in terms of information.[11] Branson explicitly compared the proteins with a message, and their constituents, the amino acids, with letters.[12]

The Russian-American astrophysicist George Gamow regarded the relation between the information coded in DNA and proteins as a translation problem.[13] The discrete language of DNA consists of digits—the four bases —and had to be translated into words (proteins). The question of the relationship between DNA and protein was thus defined as a linguistic problem, where a mathematical language must be translated into a nonphonetic, analog, continuous, and alphabetic language.

It is therefore clear, as Kay has underlined, that the expression of heredity and of life in terms of theoretical information resulted neither from any inner logic of DNA genetics, nor can it be seen as a consequence of the elucidation of the double helix in 1953; it was already in use before that time.[14] The pathbreaking work of Watson and Crick and the "deciphering" of the genetic code by M. Nirenberg and H. Matthaei in 1961 in no way contradicted the conception of life as text; rather, they provided even more arguments in its favor.

SYSTEMATIC FOUNDATIONS OF THE LANGUAGE ANALOGY

Given the multitude of linguistic terms in genetics, it is not surprising that from the 1960s some scholars have studied the comparison between language, DNA, and proteins more closely and tried theoretically to demonstrate a deeper analogy. Among the biologists the Russian Vadim A. Ratner and the Frenchman François Jacob were prominent, as was the work in the 1970s of the German Manfred Eigen, the 1967 Nobel Prize winner in chemistry, and later of his pupil Bernd-Olaf Küppers. The Russian-American Roman Jakobson, whose arguments were expanded by the German Wolfgang Raible in the early 1990s, took up the problem from the linguistic side.

Vadim A. Ratner

Ratner was the pioneer of deeper analogizing between language and genetic processes, with his work *Molekulargenetische Steuerungssysteme* (first published in Russian in 1966, then in German translation in 1977). He founded his considerations on Chomsky's conception of language as "a countable quantity (number) of sentences or sign-structures of finite length, which generate

	ANALOGUE GENETIC UNIT				
	level of nucleid acids			level of proteins	
LINGUISTIC UNIT	Jakobson/ Ratner	Eigen	Raible	Raible	Eigen/ Winkler
letter; phoneme	base	nucleotide	nucleotide		amino acid
morphe	base position				
moneme		codon			
word	codon	gene	codon	amino acid	active center of a protein
verb					enzyme
word class			homeotic gene		
sentence	gene (Cistron)	gene (simple sentence) operon (composed sentence)			proteins
paragraph		replicon/ chromosome			
whole text	segregon (=chromosome)	genome	genome	forms of the living	organism
whole text and commentary		genotype			

FIGURE 3.1 Overview of parallels between linguistic and genetic units by several authors.

an infinite quantity (number) of elements with the help of connecting operations."[15] Ratner defined "genetic language" as a "number of coding states of genetic storage," distinguishing a language of polynucleotides from a language of polypeptides. This division he compared with the dichotomy between spoken and written language. "Genetic language" would, like human language, be formed according to the principle of a *linear* concatenation of signs and be marked by a *hierarchical* structure as well as an independent grammar. On the level of polypeptide language Ratner dubbed the amino acids the alphabet and distinguished various hierarchical levels, without making explicit analogies with linguistic units. He thus left unexplained the ostensibly linguistic character of proteins. Ratner drew particular parallels only on the level of the nucleotide language. Here he alleged a linearity in the form of the phosphodiester bonds, distinguishing at least six hierarchical units: codons, cistrons (= genes), scriptons (= operons), replicons, segregons (= chromosomes), and the genome (see Figure 3.1).[16] Still, Ratner's analogizing between human language and nucleotide language treated only external structural characteristics at the expense of possible commonalities in their respective "inner structures," that is, he did not draw comparisons at the level of functions or mechanisms. Consequently, the adoption of grammatical terms for the field of genetics remained vague. In general it was assumed that the linguistic criteria of linearity and hierarchical structure introduced by Chomsky obtained only for the structural aspects of language, while leaving untouched essential qualities, like the realm of semantics.[17] As a result, Ratner came up short, with a linguistic terminology that caught only parts of language's aspects. Furthermore, hierarchical structure appears not exclusively to be a peculiarity of language, but rather a more fundamental principle to be found in all realms, living and nonliving.[18]

Roman Jakobson

The biologist François Jacob and the linguist Roman Jakobson appeared together in public in the 1960s and established, in a later-famous conversation on French television (*Vivre et parler*, 1968), commonalities between human language and DNA that were to be of great importance both to biology and to linguistics.[19] Jakobson thus legitimated in retrospect from a linguistic viewpoint the use of linguistic terms in biology. He principally cited the following linguistic characteristics as the basis of a comparison between language and DNA: hierarchical structure, double articulation, linearity, context-dependence, and redundancy.[20]

Hierarchical structure was, according to Jakobson, "the fundamental integrating principle of verbal and genetic messages."[21] The parallels he made at the level of nucleotide language corresponded with and extended Ratner's; hence Jakobson—starting from the primacy of spoken language—described at the first level not letters, but phonemes. This is important since it led Jakobson "to the most exceptional, most important and most closest analogy," *double articulation*.[22] How seriously he took this comparison between language and DNA became obvious in his parallels between codons and words: "It is quite curious that we have quite a lot of languages where the root is exactly a triplet. . . . There are structural laws of Indo-European or Semitic origin, which very much resemble this type."[23]

Jakobson's term *double articulation* was rooted in Saussure's sign-character of language and was coined and defined especially by Martinet.[24] For Martinet the human linguistic system of signs showed double articulation: a limited number of phonemes (the smallest phonetic units that are capable of conveying a distinction in meaning) can form a large inventory of morphemes (units that carry meaning).[25] Jakobson established double articulation as a criterion distinguishing human language from all sorts of other messages produced by animals.[26] Outside of human language, the only place he saw the principle of double articulation at work was in the genetic code.

Linearity represented for Jacob and Jakobson the second most important analogy between the human language and the language of polynucleotides or polypeptides. The notion of linearity equally goes back to Saussure. For him linguistic signs were bound to the linear flow of time: one sound can only follow another, thus forming a chain of sounds.[27] Similarly, in the case of DNA it is possible to speak of linearity because genetic information is stored in the linear succession of nucleotides, that is, the sequences of bases. However, Jacob had already pointed out that each occurrence of linearity required particular justification.[28] For him linearity in human language had its causes in the articulative and auditory apparatus of human beings, whereas linearity in genetic messages was due to the necessity of reproduction.

Jakobson's third criterion was *context dependence*. He referred to Clark and Marcker, who argued that codons can take on different meanings depending on their position in the genetic message. That is, they code for another amino acid.[29] In linguistics *context* refers either to the text surrounding a linguistic unit or its situative context, that is, the context of thought or sense that determines the content of an expression or the context of circumstances that determines its meaning. Jakobson clearly refers to the first. However, in order to justify an analogy between language and the genetic realm, surrounding codons would have to have an influence on which amino acid, a codon, actually codes for. This is not true from a biological point of view. When Jakobson described context dependence, he referred to starter codons. At the beginning of the coding of a polypeptide the codons AUG and GUG code for formyl-methionine, while in the middle of the polypeptide they code for normal methionine or valine. It seems, however, far-fetched to call this context dependence. Furthermore, the genetic code is universal in that even in different organisms the same codons do not code different amino acids.[30]

Redundancy was the fourth criterion for Jakobson and Jacob, a phenomenon that according to Jacob lends "some flexibility to heredity."[31] By this Jacob meant a certain synonymy in the genetic code, in that sixty-four codons code for only twenty amino acids and therefore that different codons can represent the same amino acids. In 1949 Claude E. Shannon introduced the term *redundancy* into the information-technical world in his book *Mathematical Theory of Communication*. For him, redundancy was "the fraction of the structure of the message, which is determined not by the free choice of the sender, but rather by the accepted statistical rules governing the use of the symbols in question. . . . This fraction of the message is unnecessary (and hence repetitive or redundant) in the sense that if it were missing the message would still be essentially complete, or at least could be completed."[32]

Manfred Eigen and Ruthild Winkler

In the 1970s, Manfred Eigen and Ruthild Winkler extended Ratner's approach and tried to prove structural similarities between "genetic molecular language" and human language.[33] Like Ratner, Eigen distinguished between a language of nucleic acids and a protein language, using Chomsky's *hierarchical* and *generative* principles. Whereas Chomsky's principle of hierarchy is a universal phenomenon that transcends language, Eigen saw the necessity to propose restrictions for Chomsky's generative principle in the genetic realm. The generative character of language recedes into the background at the level of the individual organism, although the evolutionary development of a living organism in its totality represents a generative process.[34]

Eigen nevertheless draws deep analogies between human language and protein language (for the level of language nucleic acids, see Figure 3.1 page 52). For him protein molecules are the smallest autonomous products of the translation of genetic messages. By equating amino acids with the alphabet or functional groups of amino acids with "actual letter-symbols," Eigen draws analogies between proteins and sentences, or between proteins and words. In support of the first, he invokes the high information content in proteins, since even the smallest consist of at least one hundred amino acids; whereas a generative principle could not work for the second, as words consist only of a few letters or syllables, and the attribution of sense would rely exclusively on "agreements made once and for all." In this case a complex representation of real circumstances by means of generative processes would become possible only at the level of sentences.[35] An argument in favor of equating proteins with words is that only the approximately four to eight active groups assembled in the active center function as symbols. This number would correspond approximately to the number of letters in a word. Eigen and Winkler thus define the active center as "the actual three-dimensional counterpart of words in this protein language."[36] Eigen abjures in this context the linguistic criterion of *linearity*; in distinction to human language, these symbols are not strung together in a line, but rather are arranged according to their chemical task in a certain spatial configuration.[37] This gives the impression that Eigen weakens the notion of language to make it conform to his intentions.

Since enzymes are responsible for executive functions in the human body, Eigen defines them as the verbs of molecular language, as words that describe actions.[38] In comparing verbs with enzymes or functional protein groups with words, Eigen clearly goes beyond Chomsky's principles of hierarchy formation and generation and projects the inner structure of language to the molecular realm. Here Eigen's analogies contradict his own observation that an analogy "is meaningful only as long as the emphasis on parallels does not obscure the differences that result from their different functions."[39]

In order to prove commonalities between the genetic and linguistic realm, Eigen introduces his own definition of language, listing the following functions of language: means of organization; means of exchange of information; means of communication; means of reflecting structures of reality by rendering possible the reproduction of complex circumstances with reductive representation; means of composition and gestalt formation; and means of self-communication (and thus the substrate of learning and thinking). Not only human language but also genetic language fulfilled these criteria.[40]

Eigen's definition of language relies strongly on his biological comparison. This becomes evident with respect to the notion of *reproduction*. Linguistic reproduction conveys extralinguistic facts by means of a medium independent of these facts; this cannot be said for genetics. Concerning the other functions, Eigen hints indeed at differences rather than at commonalities, as in his criterion of *communication*: "This communication between chromosomes and organism is unidirectional and amounts to an issuing of commands. Molecules speak with each other only at the phenotypic level, if at all, and here they use an 'object-language' oriented to optimal functional criteria."[41]

Eigen solved this contradiction by conceiving "language" as a general term for human language, protein language, and nucleic acid language. For Eigen the "phenomenon of language [transcends] the human and can be analyzed as such—thus without relation to human existence."[42] Nevertheless —and this shows the difficulty of creating a notion of language that transcends the human—he referred continually to human language, as when he wrote that on the level of transcription, "speaking," "communicating," "reading," and "understanding" simply mean "the binding (recognizing) of the right complementary molecular building blocks (linguistic symbols) and, on the basis of information supplied, the linking together of these blocks in a macromolecular ribbon (sentence), or when in relation to genetic information he drew analogies between the notion of *readability* and *reproducibility* or of *understanding* with "the ability to differentiate the organism."[43] Still, Eigen himself pointed out the differences between the genetic realm and human language: molecular genetics obeys different laws and pursues different aims from human language. He emphasized in this context that ultimately it was "irrelevant, whether the 'linguistic unit' protein represents a word or a sentence."[44]

What then are Eigen's motivations for drawing his analogies? Eigen's motivation, elucidating his attempt to find an understanding of language beyond the human, is how life and genetic information can come into being. He hypothesizes that the origins of life lie in a natural process of self-organization, and that language too derives from a process of self-organization taking place in the human brain.[45] Language, because it is a means for "self-communication," makes self-organizing evolutionary processes (genetic evolution and cultural development) possible. In the same way that human language is the substrate of thinking and acting, the evolutionary self-organization of life can be regarded as "Nature's gigantic learning process."[46] In language, as well as in the realm of DNA molecules, the gain in information is a process of learning, the "method of scientific search of truth," as Eigen calls it, bound to a dialectic proceeding. In nature as well, there is the synthesis of two modified theses in the form of genetic recombination and the subsequent reduction in meiosis.[47] The development of life is for Eigen an increase in information. However, for him the notion of information is bound to language, when he defines information as "an abstraction of a form or gestalt, as its representation in linguistic symbols."[48] The existence of a language is thus the basis for the exchange and the creation of information.

Wolfgang Raible

Raible revised Jakobson's and Eigen's analogies by incorporating newer research results from molecular genetics, using and partly modifying Jakobson's terms.[49] He also added to the tally of postulated similarities with the notions of *metacommunication* and *recursivity*.

Besides the *hierarchical* structure of the levels phoneme-word-sentence-text, Raible described the hierarchical process the recipient of a text has to work through in order to understand it. For multidimensional understanding of a *chaîne parlée* one needs to put pieces of information in greater complexes and then form a larger whole from these. The latter—defined by Raible as a "regulated combinatoric" within the hierarchy—is possible due to *metacommunicative* signals, such as different word classes (parts of speech), congruence, and the grammatical gender. Raible defines them as "signals or instructions, which orient the listener or reader and whose role is only . . . to facilitate the connection of ideas . . . to find the elements that belong together and to order them in larger units of meaning."[50] According to Raible, metacommunicative signals have nothing or little to do with an actual message; however, they provide important information about how the singular units of meaning need to be arranged in relation to each other. Raible claims that hierarchy formation and metacommunication are fundamental mechanisms also of morphogenesis, drawing analogies between homeotic genes or the homeobox and word classes.[51] This analogy with metacommunicative signals lies in his understanding of the homeobox. He writes, "The function of this homeobox may be twofold: the first corresponding to a word class in a linguistic system. It is something like a feature (sign), which allows those genes that are activated and 'read' in a very early stage of development of morphogenesis to be recognized as the member of a certain 'form class.'"[52] As a second function of the homeobox he mentions that it codes for a homeodomain protein, a "classic case of metacommunication—namely, instructions for how to convert the linear message of the genetic punched tape."[53] For Raible other transcription factors like the "leucine zipper" or the "zinc finger class" are equally to be regarded as metacommunicative signals.[54]

To define the homeobox as the "sign of recognition" of homeotic genes corresponds to a strongly anthropocentric view.[55] It is not justified from a biological point of view. To an outside observer, like a geneticist, the homeobox appears to be a sign of form classes of homeotic genes; however, during transcription inside the genetic system the homeobox is not a feature (sign) that allows genes to be recognized as members of a certain form class, but rather part of a protein-coding sequence (an exon). Whether a gene is expressed or not will be decided not at the exon, but in the control regions, which are either adjunct to the regions containing introns and exons, or located on another part of the DNA.

Metacommunicative signals have the task of transmitting the sense of information and of facilitating the receiver's understanding. This poses the question of who in the genetic realm is the receiver and what can be understood as "sense."[56] For Raible, the phenotype (the fully differentiated organism) is in the end "the message that needs to be transmitted."[57] The notion of *message* he uses synonymously with *sense*: "In the case of morphogenesis the 'phenotype,' which developed out of the 'reading' of the genome, *is* the message. In the case of linguistic communication, however, the receiver needs to *extract* the message ('sense') *from the text*."[58] Enabling the latter, however, is exactly the task of metacommunicative signals, so that in those instances where the latter need not be performed, one can only speak about metacommunication in a metaphorical sense. Raible describes homeodomain proteins and regulating proteins or transcription factors in general as "higher level instructions," where "the task of metacommunicative instructions consists primarily of . . . ones for the highly selective transformation of the complete information."[59] However, regulating proteins do not establish a meaningful context; they do not serve as aides to understanding. They might rather be compared—to stay within the metaphor—to bookmarks, meaning they regulate which part of the DNA will be expressed. To define this as metacommunicative thus cannot be justified.

Raible discusses *recursivity* as a special form of hierarchy formation, defining it as the "application of principle of gestalt formation not only in one place, but also in its repetitions at subordinate positions."[60] This principle is demonstrated in linguistics, for example, in the interlocking of one sentence in another or in the formation of a nominal group, where one or several nouns depend on another or many other nouns, as in composite nouns. Expanding on the observation that ferns, for example, form recursive structures, Raible, referring to homology and RNA processing, comments that recursivity belongs to "strategies of genesis."[61] For example, at one level similar or structurally related (meaning homologous) structures are formed in several parts of the organism in the process of morphogenesis, at another, the same DNA sequence can lead to the formation of different proteins by different processing of the pre-mRNA in different cells in different parts of the body.

Recursivity and *recursion* are mathematical notions that Chomsky applied to the linguistic realm. Beginning with the observation that the number of possible utterances in a language is literally infinite, but its grammar consists of only a limited number of rules operating on a finite vocabulary, Chomsky concluded that the same rules must be used several times when sentences are generated.[62] Recursivity is indeed a universal phenomenon, as it shows in fractal pattern formation; it is an economical phenomenon as well creating complex variety—as for example the human brain—out of a few elements; and it is an important creative principle, which applies to areas beyond linguistics and information transmission.

Following Jakobson, Raible explains the notion of *double articulation* for language as well as for the genetic code: "Just as combinations of sounds stand for something, having nothing to do with their quality as sounds, so the triplets (the genetic words) stand for something, having nothing to do with their quality as nucleotides, namely amino acids."[63] The table in Figure 3.2 provides an overview of Raible's analogies.[64]

The table shows that one can speak of a sort of double articulation in the case of DNA. But, how far does this analogy take us? Martinet's conception of the double articulation is based on Saussure's definition of the linguistic sign. Therefore I will explore how far codons show a sign character. For Saussure one does not connect the thing with a name, but rather the concept of a thing with an acoustic image. There is no material connection between thing and sign. This becomes even more evident, once the British linguists C. K. Ogden and I. A. Richards transformed Saussure's dyadic model into a triadic

	language	*biology*
first articulation	limited inventory of phonemes; phonemes alone do not mean anything	limited inventory of four nucleotides, a nucleotide alone does not signify anything
second articulation	the combination of phonemes to words corresponds to meaning	the combination of three nucleotides to a codon or word signifies an amino acid

FIGURE 3.2   The principle of double articulation in molecular genetics and language. Slightly altered from Wolfgang Raible, "Sprachliche Texte—genetische Texte: Sprachwissenschaft und molekulare Genetik," *Sitzungsberichte der Heidelberger Akademie der Wissenschaften/Philosophisch-historische Klasse: Bericht* (1993): 1–66, 60.

model, introducing "referent" thus extralinguistic reality, as a third factor. Language makes it possible to make claims about "things," not by concretely *showing* things, but by *referring* to them by means of language. Language and the outside world thus have no natural relationship; rather, between them lies a psychological mechanism. Similarly, a codon, for Raible a "genetic word," seems to stand for something else: an amino acid. However, nothing refers to anything. Rather, codon, anticodon, tRNA, and amino acid have a direct *material* relation, where tRNA functions as an adaptor molecule. So Eco writes, "'Translation' and 'transcription' are metaphors; as a matter of fact, the elements in play are coupled together because of a *stereochemical complementarity*, for the same reasons (so to speak) for which a given key fits a given keyhole."[65] Eco remarks quite perspicaciously that in this case there is indeed *substitution* like between Morse code and the Latin alphabet, whereas there is no substitution between phonemes or the units they form, the words, and reality outside language. Eco correctly describes "transcription" and "translation" as processes of "steric stimuli," where stimuli trigger or create a blind reaction, while signs always require an interpretation.[66]

Raible hints at this observation by distinguishing between "meaning" [*Bedeutung*] and "signification" [*Bezeichnung*]. Whereas Jacob wrote that each triplet has a meaning ("Each of the different triplets 'means' ['signifie'] a single unit of a protein"[67]), Raible points out that one cannot speak here in the linguistic sense of *meaning*. "Meaning" [*Bedeutung*] might be defined as "the content of a sign in a single language" and "signification" [*Bezeichnung*] by the "relation to a fact or circumstance external to language."[68] Words with different meaning might nevertheless refer to the same person or object. The fact that words bear meaning is constitutive for human languages. Here lie the roots of polysemy, metaphor formation, and other phenomena that make possible what Raible sees as the most important difference between language and DNA: linguistic creativity.[69] Codons are not meaning bearers in a linguistic sense. Therefore Raible replaces the notion of "meaning" by "signifying" claiming by this that between codon and amino acid there is a one-to-one relationship. However, the notion of "signification" refers to objects of a reality outside language, that is the notion of "meaning" as well as the notion of "signification" are only valid where there is a "intermediate mental instance."

With respect to the criterion of *linearity* Raible assumes that the coding of genetic information in the "punched tape of the DNA" and the coding of linguistic information in the linear *chaîne parlée* "solve similar problems: the projection of the multidimensional onto one dimension and the reconstruction of the multidimensional out of such linearly ordered units of information."[70] Raible postulates therefore that one should expect similar solutions to similar problems.[71] As already shown in the case of Jacob, however, the phenomenon of linearity has different causes in the linguistic and the genetic realms. Raible begins not with the problem but with the solution, that is, he errs by assuming that similar structures necessarily emerge from similar original problems. Furthermore, it is evident that in neither phylogenesis nor ontogenesis is the multidimensional projected onto the one-dimensional.

Like Jakobson, Raible describes *context dependence* in the genetic realm. For Raible, context is provided by neighboring cells in a tissue whose specific protein products can function as metacommunicative signals.[72] Raible provides a particular sentence to exemplify context: he mentions, on the one hand, the importance of adverbs for situating the sentence in time; on the other hand, he argues that "different pieces of information become unequivocal only by their coordination with others."[73] Here, however, he confounds the notions of *context* and *metacommunication*. In linguistics, metacommunicative signals make it possible to understand a sentence in its *signifying* function without making clear what is ultimately meant by it; this can only be deduced from the context.[74] In the genetic realm one cannot speak of meaningful units in a linguistic sense, and therefore one cannot distinguish between context and metacommunication.[75]

Jakobson's phenomenon of *redundancy* also figures for Raible as a commonality of language and genetics. Raible strengthens this claim with research on *Drosophila* embryos: thus for example a gene whose product regulates early stages of morphogenesis by forming a concentration gradient can be artificially doubled or tripled. If this mutated sequence is introduced into a zygote then the excess gene products provoke a change in the concentration gradient, which presumably disturbs orderly embryogenesis, leading to severe damage in the organism. However, it was observed that—because of the phenomenon of redundancy—the consequences of this mutation can be counteracted leading to a normal organism. Here redundancy can appear in very different forms: first, one and the same position can be occupied by different gene products, which exercise the same or different effects; second, it is possible that one and the same result is provoked by different gene products. Redundancy in the genetic system means that the same process, for example gene expression, can be triggered in different ways or by the different interplay of various factors. Redundancy makes sure that the right information is used in the right place and at the right time. Even slightly modified sequences in an amino acid protein will not necessarily affect its function. However, the cell performs no mental process of—in Bühler's phrase—apperceptive complement. The connection is purely material, in the way that even a defective key can open a lock. Since redundancy serves to secure information, this indicates the function of DNA in information storage.

ANALOGIES ON THE BASIS OF MODELS OF COMMUNICATION

Chomsky's formal language theory was central to a large part of the work on analogies between language and molecular genetics. Indeed the comparison is most convincing where it emerges from a mathematical understanding of language. Since Chomsky's conception of language excludes semantics and ignores the communicational aspect of language, it seems worthwhile to explore how far the analogy goes with other concepts of language. In the following I will discuss briefly the well-known models of communication by the German-American psychologist Karl Bühler and Jakobson, and second the work on language universals of the American linguist Charles F. Hockett. Bühler's organon-model and its extension by Jakobson render the term language more precisely:

1. A speech act requires at least two individuals with language mediating between sender and receiver.

2. The sender can become a receiver and vice versa.

3. Language is social.

4. Language is both objective and subjective: it is objective, because there is a code that stands above sender and receiver, and subjective, because language is actualized only in the individual.

5. Language requires a means of transmission.

6. Linguistic signs are bearers of meaning.

7. Language is reflexive; that is, with language one can talk about language.

8. Language is poetic, or can be poetic.

9. Language is a mental and a psychological function.[76]

Some of these criteria can be rapidly dealt with for the genetic realm: there is no mental instance and therefore there are neither reflexive nor poetic functions. As shown, one cannot speak of signs in the genetic realm. DNA could function as a means of transmission; however, here the information is conveyed in a biochemical manner, while human language relies on physical regularities. The properties "social," "objective," and "subjective" relate to the position of sender and receiver. Various authors have tried to solve this issue differently. Eigen, for example, avoided the notions of sender and receiver because of the vectorial character of DNA-information.[77] Raible's definitions can be summarized in the table in Figure 3.3.

Küppers regards the genome as a *sender* and the cell as the *receiver* of biological information.[78] The mRNA, that is, the modified complementary copy of the DNA, probably figures as text.

The following principal differences from human language are clear from these comparisons: (1) the receiver cannot become the sender, or vice versa; and (2) sender and receiver have different biological compositions; that is, while two human interlocutors are individually different, they are nevertheless similar in their human constitution (*Bauplan*) whereas the molecular sender and receiver are qualitatively different. In Küppers's comparison, however, sender and text have similar material constitution, even though mRNA

	sense	sender	text	receiver	sense
1. *level of reproduction*	adult	gametogenesis (sending)	zygote	development (reception)	adult
2. *level of transcription and translation*		copy of the DNA (producer)	text of the DNA	RNA-polymerase and ribosomes (recipient)	
3. *level of the cell*			genome, text of the DNA	cell (reader)	

FIGURE 3.3   Overview of parallels between the genetic realm and language, which are based on communication processes. The terms in brackets are Raible's.

is complementary to the DNA. Transmission thus has no proper medium; the use of the term *communication* in relation to the genetic realm is therefore exclusively metaphorical. The apparent paradox that one can speak quite correctly of the information storing DNA as a text, can be solved—as Eigen and also Raible imply—if the word "text" is replaced by "program."[79]

With the aim of delineating human from animal language Hockett listed sixteen design features of adult human language, ten of which he considered so basic that he used them for a general characterization of languages: "Any system that has these ten properties will here be called a language; any language manifested by our own species will be called a human language."[80] In summary, they are:

1. Vocal-auditory channel: language is vocal-auditory (written language is merely derived).

2. Rapid fading: related to this point, linguistic signals are evanescent.

3. Interchangeability (the sender can become receiver and vice versa).

4. Complete feedback: the transmitter of a linguistic signal himself receives the message.

5. Specialization: the direct-energetic consequences of linguistic signals are usually biologically trivial; only triggering effects are important.

6. Arbitrariness: the relation between a meaningful element in language and its denotation is independent of any physical or geometrical resemblance between the two.

7. Discreteness: the possible messages in any language constitute a discrete repertory rather than a continuous one.

8. Displacement: linguistic remarks may refer to the past or to the future.

9. Openness: a finite repertoire of signs and grammatical rules can produce an infinite number of messages; this corresponds to the notion of creativity.

10. Duality: this characteristic corresponds to Martinet's double articulation.

Hockett pointed out that language is acoustic and therefore fades rapidly.[81] Raible also saw the latter as a difference from the genetic realm.[82] The former is obvious as attempts to draw analogies with human language tend to refer to the written form. "Interchangeability" and "complete feedback" are not valid in the genetic realm, where—as already argued—sender and receiver cannot be unequivocally differentiated. For specialization Hockett provides an example that even a passionate discussion cannot increase the temperature in a room and no external physical factor can influence language *directly*. However, in the genetic realm, physical factors can lead to severe disturbances of the transmission of information; DNA can denature with increasing heat, for example. Arbitrariness and discreteness relate to the sign character of language. Of course DNA has discrete units (bases of the DNA, triplets), but as I have shown these have no sign character, that is, they have no denotation and are not bearers of meaning in the genetic realm. In the genetic realm past and future cannot be expressed—corresponding to Hockett's displacement—although there is a historic dimension, which—to remain in the metaphor of language—one could call *Sprachwandel* (change of language).[83]

SUMMARY OF THE COMPARISONS

Each of the authors discussed earlier defined linguistic units slightly differently and introduced new terms (like Eigen's "moneme" or Ratner's "morphe"). Nevertheless, Figure 3.1 and the subsequent discussion show clearly that the different analogies of linguistic with genetic units are not the result of different linguistic conceptions, but rather of the fact that genetic units are interpreted differently. It becomes clear that there are no unequivocal parallels, but that the attributions reveal strong subjective judgments. Some authors refer only to structural similarities: smaller genetic units are equated with smaller linguistic units, and larger genetic units to larger linguistic units. The principle on which this parallelism is based is thus purely hierarchical by size. However, hierarchy by itself is not a specific linguistic phenomenon. Other comparisons refer to functional aspects (verbs, word classes). However, where parallels might hint at actual essential kinship, the comparison breaks down and appears artificial. Furthermore, the comparisons are not absolutely consistent. There is a clear break in argumentation in Raible's analogy on the level of nucleic acids: on the one hand, he equates the codon with the word; on the other, he asserts that a homeotic gene represents a word class. If the latter is true, then it must be the gene—and not the codon—that corresponds to a word. Eigen's and Winkler's analogy goes into great detail on the level of words, but remains vague and without exact correspondences on the higher levels. In general, it appears that these authors draw their parallels quite arbitrarily and establish comparisons on the flimsiest of pretexts.

Ratner, the pioneer of a deeper analogy between language and genetic events, did not intend to establish or justify a consistent, stringent analogy between human language and the genetic realm. Instead "linguistics"—or better "written language"—was for him a "sufficiently good analogon" to analyze and describe the structure and function of the genetic system. Deeper analogies Ratner found only on some levels, as analogies with language were not the central aim of his work, but only a means to an end.

Jakobson took the comparison so seriously that he derived claims about the origins of human language from it. He advanced a theory that the origins of language are based on the principles of molecular genetics: "I believe that it is not too daring to suppose that this structure, this similarity of structure between molecules and language is due to the fact that language was, in its architecture, modeled on the principles of molecular genetics, because it is as much a biological phenomenon as this structure of language [*langue*]."[84]

Jacob, in his capacity as a biologist, argued against this: "For the biologist, in effect, it is difficult to see how the structure of human language could have been modeled on that of heredity."[85] Jacob defended the idea that the analogy results from similar functions, which emerged in convergent development.

Eigen's parallels, based on Chomsky's and his own definition of language, lack rigor and have little basis in linguistics. His definition of a language independent of human language is not convincing: first, it is defined by his comparisons to genetics; second, it is modified according to circumstances; and third, Eigen has continual recourse to the properties of human language to justify his comparisons. Still, Eigen is of great interest, because he emphasizes the creation, transmission, and storage of information as central both to human language and the genetic realm. His work shows the difficulty of getting a grasp on any notion of information independently of language. Still, further study of the notion of information may hint at possible analogies.

Unlike Eigen, Raible makes his analogies explicitly with human language. His criteria (as developed previously) try to explain the "functioning of linguistic communication" and to show that "'the grammar of biology' shows surprising similarities to structural properties of the grammar of human languages."[86] Raible uses the term *communication* in a misleading way, since he does not provide a model of communication, but only introduces those criteria that concern the realm of understanding spoken and written language. As Raible himself says, he describes in fact only the "problem of meaning synthesis."[87] Raible's arguments are weakened by the absence of a definition of language and its functions, the introduction of the term "communication," and his explicit comparison with the grammar of human language. In general the characteristics he discusses are hardly sufficient to explain the grammar of human language.

However, it is striking that almost all of Raible's terms have significance in computer science, which makes a comparison to *artificial* languages possibly more plausible. What is understood here by *context-dependence* (the milieu, in which stored information becomes operational) or by *reading* has its parallel in computing, since software cannot be treated without hardware. Computer programs show a linear sequence of bits, are structured hierarchically and recursively, and have redundant sequences. What Raible understands by metacommunication also refers to the construction of loops in computer programs: the computer can jump to a certain place in the program and call up the information stored there. Another example is Hockett's term "discreteness" (that information needs to be unequivocal), which was true for linguistic units as well as for codons.

An exception is *double articulation*, which does not occur in the genetic realm, because the notions of *meaning* and *signification* do not apply without an intermediate mental instance. Therefore on the level of transcription and translation one must speak about—to use Eco's term—"substitutions." Thus the discussion around double articulation makes it necessary to refine the terminology involved and to look more closely at the misuses of the term *meaning*. This will ultimately bring me back to the field of *information*.

In 1968 the French geneticist Philippe L'Héritier, in a discussion with François Jacob, Roman Jakobson, and the anthropologist Claude Lévi-Strauss, raised the question: "What does it mean to say that an organism has a meaning?"[88] Such a question is especially relevant for Küppers who does not distinguish between "meaning" and "sense." Referring to Darwin's evolutionary theory, Küppers writes that "the purpose-directed, information-controlled, element of biological structures possesses a certain function . . . with regard to the preservation of the dynamic order that is characteristic of living systems," thereby assigning to this function 'sense' and 'meaning.'"[89] This sense of "meaning," however, is "characterized solely by the function (of information)."[90] Stenglin, who has worked on the treatment in the nervous system, points out: "The equalization of function and meaning of information is here valid for mechanical processes: that is, either automated or non-spontaneous physiological processes, similar to computer programs, which execute a large number of if-then instructions."[91] Thus "meaning" and "function" converge.

Therefore the question raised by L'Héritier should be answered in the sense that one needs to distinguish the notion of *meaning* from the one of *sense* on the level of the text, as Oesterreicher did: "The special character of linguistic content, expressed in texts, should—in distinction to *meaning* and *signification*—be called *sense*."[92] Whereas "meaning is not explained from outside (authority above language), but only from within language," the *sense* of a system, a text, and so on will become clear with an external analysis.[93] DNA is clearly a bearer of *sense*: from an ontological point of view it is the construction of the organism; from a phylogenetic point of view the survival of the species under changing environmental conditions. The notion of *meaning* is clearly connected to human language and mental processes, whereas in general the realization of *sense* is the aim of information-treating processes.

Clearly, the transmission and storage of information is a key commonality between language and the genetic realm in almost all of these works. Whereas Eigen started with the notion of information and saw the necessity of dealing with the notion of language in order to get a grasp on "information," other authors started with the notion of language and developed criteria that led to the notion of information. The notion of information is thus at the intersection between language and the genetic realm.

I have shown in this work, on the one hand, how fast the analogy with human language comes to its limits, and on the other, how it is impossible to grasp the notion of information independently of language. Küppers attributes a "defined semantics" to information, but in the end it lacks precision, as meaning and sense converge; Eigen gives an unconvincing abstract notion of language; and Weizsäcker addresses openly the difficulties of a definition of information based on one of language: "What do I mean, when I say that information has linguistic character? I have not even first defined what language is."[94] Neither for language nor for information is there an unequivocal definition, but, since one depends on the other, this is necessary for any further explanation.[95]

THE USE OF LINGUISTIC PROCEDURES IN
THE GENETIC REALM

Shortly after the adoption of information-theoretical terms in genetics linguistic procedures were applied to DNA. The astrophysicist George Gamow conceived of the connection between the structure of DNA and protein synthesis as a "mathematical cryptanalytic problem" and gave it to navy specialists, who had worked on the decipherment of enemy codes. After two weeks they gave it back as insoluble. Other leading scientists in the 1950s also tried to solve the code by cryptographic and mathematical means, without coming any closer to a solution, impeded by linguistic metaphors. All natural languages show "intersymbol restrictions," so that the statistical analysis of an unknown code is based on nonaccidental distribution of the frequency of letters and neighbor relations. In distinction to this, the genetic code is neither restrictive nor overlapping. Also, the application of Shannon's theory of information with the assumptions of a binary code, as tried by Robert Sinsheimer, did not lead to a result.[96]

Beside the coding sequences there are noncoding sequences, whose function is still in the dark. As in the early phases of molecular genetics, biologists are once again trying to find solutions with linguistic and information-theoretical procedures. Thus Mantegna et al. used the Zipf test (devised by the American linguist George Zipf) for DNA and a redundancy test based on Shannon's information theory.[97] Zipf had counted words in texts of different languages and had listed them according to the frequency of their appearance. He charted the logarithm of the rank of a word against the logarithm of frequency and obtained a linear function for all texts.[98] Why this is so, is a mystery: "There's no rhyme or reason why this should be true."[99]

The redundancy test relies on that a random series shows no redundancy, meaning that "the shortest algorithm needed to generate the pattern of this sequence is of about the same length as the sequence itself."[100] In order to use both tests for DNA, it was necessary to define what corresponded to a word on a known base-sequence. In relation to the coding sequences a word was equated with a triplet (n = 3 bases). Since for the noncoding DNA sequences no functional sequence is known, "words" were created arbitrarily (each between n = 3 and n = 8 bases). The various n-groups were obtained by moving one base further on a given DNA sequence. DNA-sections from different organisms were tested with the Zipf test; although the test failed for the coding sequences, it showed a linear function for the noncoding sequences. As the "words" had been created arbitrarily, the Zipf test was applied in a check-test to a collection of texts, which came out of an encyclopedia, as well with natural words at the basis as with arbitrary n-groups. In both cases the Zipf test was positive, although with a higher index for the first case. Also the binary series of a computer program of nine millions bits was examined with the Zipf test, which ended with a positive result, whereas the Zipf index was negative when an arbitrary series was chosen. The redundancy test confirmed the results obtained in the Zipf test. Not only the computer program and the human language contained redundant sequences, but also the noncoding section of the DNA (junk DNA). There was no redundancy for the coding DNA.

The researchers were not surprised by these results: "The familiar coding regions of the genes fail both tests—an expected result, Stanley says, because the 'language' of the genes lacks key features of ordinary languages. 'The coding part has no grammar—each triplet [of bases] corresponds to an amino acid [in a protein]. There's no higher structure to it.' In contrast, junk DNA's similarity to ordinary languages may imply that it carries different messages, says Stanley."[101]

It is striking how Stanley dismisses with one sentence an analogy between coding regions of DNA and language (taken for granted in numerous earlier works), and assumes that it is rather junk DNA that has the character of language. It would be quite surprising if introns had a linguistic character, since in general exons are the bearer of genetic information, and introns are excised in the process of splicing and therefore are presumably only relics of evolution. Hedgecoe defines only the coding parts of the genes as "informational genes," opposing them to the "biochemical genes" in the complete picture of exons and introns.[102]

In Mantegna's work and the subsequent work of his group, the existence of "long-range correlations" in the noncoding DNA is seen as further proof of its linguistic character. Bonhoeffer et al. argue that the higher value for redundancy r(n) of noncoding DNA could simply be a result of its larger variance of the distribution of the frequency p_i.[103] By charting the relative frequency of words against the rank of words a human DNA sequence and an arbitrary sequence of the same length and similar frequency of nucleotides displayed similar curves. Furthermore, the higher occurrence of certain combinations of nucleotides, beyond a random variation, might have its cause in an unequal crossover. This is emphasized by Voss: "The standard analysis check, omitted from [Mantegna et al., 1994], of randomizing the sequences shows that, when observed, the differences are primarily due to nonuniform base probabilities."[104] Israeloff et al. did control studies and came to the conclusion that "to detect language Zipf analysis should be applied with caution, since it cannot distinguish language from power-law noise."[105] Furthermore, besides these objections, it is important to emphasize that the Zipf test does not provide useful information about natural language and that its relevance is unclear. Attard et al. developed a physical model to explain the origin of structures possibly similar to languages. However, the authors came to the result that "the language-like features of noncoding DNA could simply represent the footprints of mobile element insertions and excisions that have occurred over evolutionary time scales. The implication of these findings is that if noncoding DNA does indeed have some higher-order function, this function is an opportunistic evolutionary development which exploits the underlying structure of junk DNA."[106] Tsonis et al. did a test on the assumption that "real DNA sequences must be obtained if sentences were 'transformed' to DNA sequences."[107] As letters in a certain alphabet have a certain frequency distribution like codons in DNA sequences, they associated letters and codons according to their frequency. So sentences could be transferred in coding and noncoding DNA sequences (coding sequences have a different frequency distribution from noncoding ones). Thus, syntactically correct sentences as well as random series of words were transferred in DNA sequences and compared to existing DNA sequences. Syntactically correct sentences as well as the non-sense sequences showed the same homology. Also after an arbitrary rearrangement of letters in some words the same parallels were observed. The authors concluded, "The inescapable conclusion is clear: DNA sequences show no linguistic properties."[108]

Molecular linguistics today uses methods that rely on syntactical recognition of patterns and on formal language theory in order to explore the structural properties of DNA and proteins.[109] The success of this procedure seems to lie in that one departs from the outer structure of language and from general characteristics of information-treating processes, whereas the transmission of inner structures of the human language and its application of linguistic tests does deliver null or dubious results.

FINAL REMARKS

I have used a traditional concept of metaphor, according to which a metaphor is an image, an unreal use of a term. The heuristic value of a metaphor relies on the fact that we do not choose metaphors arbitrarily, but rather on the basis of perceived similarity. Natural scientists since the seventeenth century have regarded metaphors with suspicion, calling them imprecise and subjective, and thus unfit for scientific language. However, science has more recently again paid attention to metaphor's heuristic and constitutive values. There are "cases in which there are metaphors which scientists use in expressing theoretical claims for which no adequate literal paraphrase is known. Such metaphors are constitutive of the theories they express."[110]

In fact these are no longer metaphors in any strict sense: heretofore unknown processes had to be described, and in this sense the expressions used were not metaphors, as they did not replace real expression by unreal ones. What then would be the real expression for transcription? The only familiar term, which describes the production of mRNA from DNA, *is* transcription. So Rorty is right when he writes, "Any argument to the effect that our familiar use of a familiar term is incoherent, or empty, or confused, or vague, or 'merely metaphorical' is bound to be inconclusive and question-begging."[111] The expression "genetic molecular language" is thus—dependent of the point of view—real term or metaphor; it is, however, not more than a metaphor inasmuch as there is no identity between it and the human language.

The biological understanding of analogy attributes similarities in the *Baupläne* of related animal and plant species to the exercise of similar functions. In the expression "the language of the genes" metaphor and analogy complement each other in the sense that the metaphor reveals an analogy: the metaphor is the linguistic expression of a functional similarity. However, this does not concern all the functions of human language, but rather aspects of their transmission, storage, and securing of information. The observation of congruence, of identity, is—to express it metaphorically—the death of the metaphor. Jakobson, Jacob, Eigen, Küppers, and Raible see linguistic structures in the genetic realm, but they do not claim congruence or identity.

Hedgecoe argues for a "constrained constructivism" in the use of metaphor in the natural sciences, which accords with demands for scientific objectivity but also relies on cultural and personal experiences.[112] Keller writes, "A metaphor is in the first instance a vehicle, carrying meaning from one referent to another, reflecting one aspect of reality through the lens of another, and back again. As such, it also has a kind of force, forging new ways of perceiving the world, and hence, new ways of acting in and upon the objects of our perception. Similarly, because a metaphor is a way of talking or writing, and itself a communicative act, it is also a vehicle for transmitting new ways of perceiving and acting, new orderings of the world, from an author (or speaker) to his or her audience."[113] Both authors locate metaphor between subject and reality, or between subject and society, and argue that it forms our consciousness and our understanding of the world, and therefore should be used consciously and with care. Thus neither the metaphor itself, nor attempts to derive deeper analogies from it, should prevent access to new results.

In the general discussion and perhaps especially among biologists differences have been neglected. Jacob has stressed that biology has no real mathematical theories and therefore requires models. This could—as he emphasizes—lead to the model's serving for the explanation and that analogies are interpreted wrongly: "Thus the frequent tendency to take the model for an explanation and analogies for identities."[114]

Scholars' surprising persistence in applying the language metaphor to genetics demands explanation. There is, no doubt, a pragmatic element: this metaphor has come a long way in biology and assumed an immense heuristic value, especially in the analysis of mutations: "Once a genetic message was compared to a written alphabetic text, then it seems logical to equate mutations with copying errors introduced into the text by the copyist or printer."[115] However, there are also genetic metaphors that do not rely on language, like the "leucin zipper," that nevertheless usefully reflect the function of the objects under study.

Linguistic notions are helpful for preliminary understanding; they attach themselves easily to the genetic realm. Likewise, linguistic methods based on mathematics and statistics are used in genetics—although, it must be stressed with ambivalent and controversial results. Research has focused especially on the notion of *information*, at the intersection of the linguistic and the biological worlds. But this alone does not fully explain the persistent uses of the analogy. The wide use of metaphor is therefore also due to a vague conception of *language*. The answer to the question of whether the "language of genes" is more than metaphoric, depends on what we understand by "language." Thus Hedgecoe is right when he writes, "It is possible to view the language metaphor as transformed, as subject to 'full-fledged investigation . . . using discipline-specific procedures,' but this would involve a different interpretation of language."[116]

Moreover, the embeddedness of the language metaphor in Western culture certainly plays a major role, since the metaphor of the *book of nature* is rooted in the Judeo-Christian creation of the world by God's word. For many biologists, drawing parallels with the linguistic realm meant asking philosophical questions about the origins of life and the secret codes of the human *Bauplan*. In the end they optimistically hoped to answer the question of "What is life?" by finding answers to "What is language?" and vice versa. But asking these metaphysical and naturphilosophical questions of an interdisciplinary nature seems to have misled many researchers to search for greater unity and more complete explanation, mostly in flat contradiction to research and results in biology, computer science, and linguistics. Nevertheless, perhaps human beings can arrive at an exact knowledge of their own *Bauplan* and those of animal and plant species through a general knowledge of information-trading processes. We humans would then be not only the first to understand DNA in total, and, in a certain sense, to read it, but also, as Keller aptly notes, to write anew.[117]

CHAPTER FOUR

Language in Action: Genes and the Metaphor of Reading

Evelyn Fox Keller

Language has provided a set of handy metaphors for thinking about the relationship between germ cells and the organisms they give rise to, or as we would now put it, between genotype and phenotype, ever since the earliest days of genetics. August Weismann may have been the first to employ a linguistic metaphor for this relationship, invoking it, ironically, to argue for the theoretical impossibility of Lamarckian inheritance. He wrote:

> If we were now to try to think out a theoretical justification [of Lamarckian inheritance] we should require to assume that the conditions of all the parts of the body at every moment, or at least of every period of life, were reflected in the corresponding constituents of the germ-plasm and thus in the germ cells. But . . . [this] is very like supposing that an English telegram to China is there received in the Chinese language.[1]

The same metaphor recurs, toward the end of his argument, in a refutation of the more specific claim that individual memory can be transmuted into genetic memory: "I can only compare the assumption of the transmission of the results of memory-exercise to the telegraphing of a poem, which is handed in German, but at the place of arrival appears on the paper translated into Chinese."[2] We may of course question the appropriateness of any linguistic metaphors for vital processes, and especially the metaphor of translation, but the most conspicuous irony here is in the particular use to which Weismann puts the metaphor of translation. As Walter Benjamin reminds us, the very function of translation is to express "the central reciprocal relationship between languages"; and indeed, language itself is rooted in the possibility of two-way communication.[3] Yet here, Weismann employs the linguistic metaphor precisely to illustrate the essential asymmetry of the relation between germ cells and body parts (respectively figured as Chinese and either English or German speakers), that is, the impossibility of two-way communication. The germ cells may "determine" the fate of memory cells, but Weismann explicitly rejects Richard Hertwig's earlier suggestion that the process might be reversible: the converse appears to him to be simply inconceivable. He acknowledges that "no one can state with any definiteness how the germ goes to work," but nonetheless argues that, even though "the process cannot be understood in detail, it can in principle, and this is just what is impossible in regard to the communication of functional modifications to the germ."[4] In an apparent reversal of the more familiar forms of Orientalism, the germ cells, written in Chinese, are interpretable by Europeans, but the language of somatic cells (written in English or German) remains forever inscrutable to the germ cells, which are literate only in Chinese. We might also at least parenthetically note that Weismann's earlier diagrammatic representations of the relationship between germ and somatic cells had displayed considerably more nuance, allowing for a certain amount of crosstalk between germ cells and early progenitors of somatic tissue, but, as Griesemer and Wimsatt have shown, by 1904, his diagram had already been transmuted into its stronger (and more familiar) form by E. B. Wilson and his students.[5] (Weismann's original diagram of the germ-track is shown in Figure 4.1, while Wilson's redrawing of this diagram is shown in Figure 4.2.) In Wilson's redrawing of Weismann's original diagram, germ cells are depicted as sending uni-directional messages to somatic cells, and as reading only those messages sent by other germ cells.

We could of course question Weismann's claim that by 1904 an "in principle" understanding of the process of genetic determination was already at hand, but my point lies elsewhere: it is to note the singular inappropriateness of a linguistic metaphor for the particular argument Weismann wishes to make. What is it doing there, and how does it work? And more puzzling still, how are we to understand the choice of Chinese as the language into which (but not from which) translation is barred?

To ask how a metaphor works is in the first instance to ask a question about language: the metaphor of language has a built-in reflexivity; it is itself a form of communication, employed by one language speaker in an effort to persuade other speakers of the same language. In other words, by virtue of being in language, metaphors are themselves speech acts, acting in and on the world. And a metaphor for language is, at the same time, a metaphor for that act. Translation invokes both the real and the metaphorical enlargement

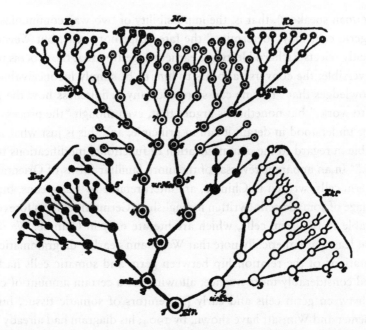

FIGURE 4.1   Weismann's Diagram of the Germ-track of Rhabditis nigrovenosa.
"The various generations of cells are indicated by Arabic numbers, the cells of the
germ-track are connected by thick lines, and the chief kinds of cells are distinguished by various markings:—the cells of the germ-track by black nuclei, those
of the mesoblast (*Mes*) by a dot in each, those of the ectoderm (*Eke*) are white,
those of the endoderm (*Ent*) black; in the primitive germ-cells (*ur Kz*) the nuclei
are white. The cells are only indicated up to the twelfth generation." A. Weismann,
The Germ-plasm: A Theory of Heredity (New York: Scribner, 1893), Figure 16, 196.

of the audience to include speakers of different languages. In Weismann's
text, the metaphor of transmission is itself a transmission, sent by the author
to his readers. Now, who are Weismann's readers? He is of course writing in
German, to an audience composed primarily of German zoologists, but by
the end of the nineteenth century, his works are appearing in English translation, and they are read by American and English zoologists as well. Notwithstanding the romance of German academics of Weismann's generation
with all things Oriental, his works do not appear in Chinese. Furthermore,
and undoubtedly more germane, it might be argued that the "hardening" of
Weismann's position in the last decade of the nineteenth century and the first
decade of the twentieth (discussed by Griesemer and Wimsatt) parallels the
bifurcation then beginning to emerge between geneticists and embryologists,

a bifurcation dividing advocates of nuclear control from those of cytoplasmic control.[6] Weismann clearly meant for his readership to include both geneticists and embryologists, but his intended successors—those who could be counted on to relay his principle of biological organization to future generations, and who could thereby shape the future of biology, were the students of germ line transmission, that is, the geneticists. Like the genes they studied, geneticists had a root message to transmit, a message that embryologists, like the somatic structures they were studying, were expected to receive but not to be able to answer. The transmission structure of Weismann's message to biologists is depicted in Figure 4.3.

Weismann was clearly ahead of his time. Neither he nor his readers could possibly have anticipated a material structure for the genetic determinants—a structure that did indeed appear to be constituted as a "text" in the literal sense of that term. This development—responsible for what might be called the golden age of the linguistic metaphor in genetics—needed to wait half a century; it needed to wait for the identification of DNA as the genetic material, and of the text of life as a sequence of nucleotide bases. The genetic

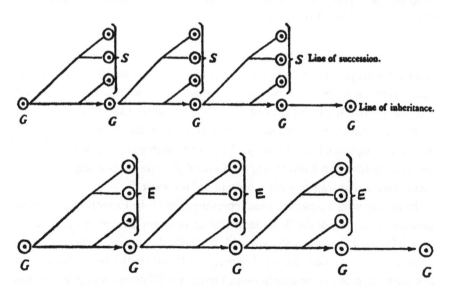

FIGURE 4.2 *(above)* Diagram illustrating Weismann's theory according to Wilson. From E. B. Wilson, *The Cell in Development and Inheritance* (New York: Macmillan Company, 1896), Figure 5, 13.

FIGURE 4.3 *(below)*   G = geneticists; E = embryologists.

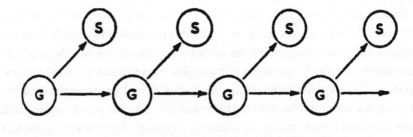

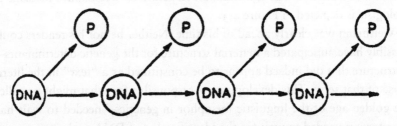

FIGURE 4.4   Maynard Smith's redrawing of the Weismann-Wilson diagram. Reprinted from J. Maynard Smith, *The Theory of Evolution*, 2d ed. (Middlesex: Penguin, 1965), Figure 8, 67, with permission of Penguin Books Ltd. London, © John Maynard Smith 1958, 1966, 1975.

message, we soon learned, is carried by the messenger (RNA) to the ribosomes, whereupon it is translated into protein. And just as Weismann had argued, so too, now Francis Crick asserted that "once 'information' has passed into the protein it cannot get out again."[7] Careful to distinguish this claim from the "sequence hypothesis," Crick dubbed it the "Central Dogma." Where the sequence hypothesis was "a positive statement," asserting the necessity of translation from DNA, the Central Dogma "was a negative statement, saying that transfers from protein did not exist."[8]

By juxtaposing a diagrammatic representation of the Central Dogma with his own redrawing of the Weismann-Wilson diagram (as seen in Figure 4.4), John Maynard Smith in 1965 was able to claim the success of molecular biology as a clear vindication of Weismann.[9] Of course, the two diagrams do not quite map onto one another—in Figure 4.2 (Wilson's redrawing of the Weismann diagram), the recipient of the genetic message is the Soma, or body, where in the new version (Figure 4.3), the body is signified simply as *P* for protein. But the point of immediate interest here is that Maynard Smith, like Crick, employs the same metaphors of language and translation as did

Weismann in support of their interdiction of reverse translation. Indeed, Maynard Smith elsewhere quotes Weismann's original analogy of translation into Chinese, and comments, "It is hard to imagine a clearer expression of the theoretical difficulty of Lamarckian inheritance."[10]

To be sure, the ground for translation metaphors provided by molecular biology was vastly more secure than any that Weismann had had access to, and by the mid-1960s, geneticists no longer had reason to regard the genetic text as inscrutable. A simple isomorphism had been demonstrated between sequences of amino acids and nucleic acids, and a manual for translation was available for the asking. But translation was still a metaphor implying reciprocity—by definition, referring to a reversible process. On what grounds, then, was the possibility of reverse translation here precluded? Surely, none that molecular biology had been able to provide. As Crick himself noted in the very paper in which he first articulated the Central Dogma, this was a hypothesis "for which proof is completely lacking."[11]

Just as with Weismann's earlier principle, Crick's message also is twofold: the Central Dogma describes a message transmitted from DNA to protein, but at the same time, it is itself a message, a founding principle ordering not the cell but the world of biological science. It is a principle written in the first instance by Crick, received and relayed by other molecular biologists, and uni-directionally transmitted to all those nonmolecular biologists who have been rendered voiceless by the phenomenal successes of molecular biology. (The transmission structure of Crick's message to the biological community is depicted in Figure 4.5.)

But just because these successes were so substantial, a third meaning of the metaphor of genes as text now comes into view. Once the code by which sequences of nucleic acids are translated into sequences of amino acids had been cracked, a new kind of reader is called into being. In addition to readers of

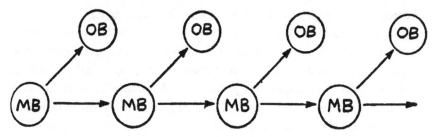

FIGURE 4.5   OB = ordinary biologists; MB = molecular biologists.

genes (that is, cells) and readers of scientific texts (other biologists), it now becomes possible to envision human readers (that is, molecular biologists) who can actually read, and decode, the genetic text itself. But interestingly, this vision of texts and readers included no implication of a barrier that might impede the new kind of reader from writing back, that is, from re-writing the text. In fact, quite the contrary. From its very earliest appearance, what seemed most exciting and compelling about such a vision (at least to molecular biologists) was precisely the power of rewriting that such reading competence promised to bring.

The prospect of directly reading DNA (and with capacity, the clear expectation of then being able to rewrite that script) loomed large in the late 1960s, but at that time it was still only a future to be glimpsed. In 1969, as he recalled the "old dream" of "the perfection of man," Robert Sinsheimer wrote, "It is a new horizon in the history of man. . . . We now glimpse another route—the chance to ease the internal strains and heal the internal flaws directly, to carry on and consciously perfect far beyond our present vision this remarkable product of two billion years of evolution."[12]

But time moved fast for molecular biology in the 1970s and 1980s, and within less than twenty years, the possibility of reaching out and actually touching—even grasping—the new horizon seemed to be at hand. The project to sequence the human genome was launched with the explicit promise of "controlling human suffering," and just beneath the surface lay the tacit promise of realizing that old dream of "perfecting man." Much has been written about the new rhetoric of reading and writing the "book of life," even of making "perfect babies," and I do not need to review that literature here. For my purposes it suffices to note the striking difference between the new readers, empowered to rewrite just by virtue of their reading skills, and the older readers—be they cells or nonmolecular biologists—who, although competent as readers, were assumed to be incompetent as writers—indeed, rendered so by the very force of the message they were given to read.

Once again, thinking about the inherent reflexivity of the linguistic metaphor is useful. With the emergence of a new kind of reader, the message itself takes on a new, a third, level of meaning. Let us again ask, who is an author, a sender of messages (like genes or DNA), who is a reader, like cells or protein molecules, capable only of receiving, and who is the reader competent to revise the message read? One of the most striking aspects of the new rhetoric of molecular biology is the range of its audience—the world of readers now includes not only other biologists but also congressmen, patients, biotech investors, ordinary people. Accordingly, the reach of the linguistic metaphor is vastly extended, working to order not only the organization of cells and of biological science but also, in a sense at least, of the larger social order.

An early indication of the new, extended, domain of ordering can be detected, for example, in some of the comments Francis Crick offered in 1963 on the subject of "Eugenics and Genetics," especially once we pay close attention to his particular use of personal pronouns. I quote three of these comments: (1) "Do people have the right to have children at all? . . . I think that if we can get across to people the idea that their children are not entirely their own business . . . , it would be an enormous step forward."[13] Two subjects are explicitly indicated: the "we" who knows, and those to whom "we" must communicate our knowledge, namely "people."

A similar syntactical distancing recurs in another, related, comment, but now further enhanced by use of the even more abstract pronoun, "one": (2) "The question . . . as to whether there is a drive for women to have children and whether [my proposal of licensing] would lead to disturbances is very relevant. I would add, however, that there are techniques by which one can inconspicuously apply social pressure and thus reduce such disturbances."[14] And for a final and truly embarrassing example: (3) "We are likely to achieve a considerable improvement [in the human stock through genetics] . . . that is by simply taking the people with the qualities we like and letting them have more children."[15]

In these early days of molecular biology, when the prospect of rewriting the genetic text remained a thing for the future, not yet realized, a simple (Central Dogma-like) structure for the relationship between author and reader still seemed to obtain. The main difference between Crick's reading paradigm and that which Weismann would have assumed is to be found in the identities of author and reader: the generic authorship Crick has in mind (his "we") consists of molecular biologists, and the passive readership for which he writes, the readership he wishes to inform, extends well beyond "other biologists" to include that much larger category of ordinary "people." (Figure 4.6 depicts the transmission structure of this enlarged community of readers.)

But in 1963, the days when so simple and so quintessentially "modern" a structure might even be thought to pertain were already numbered. Crick's

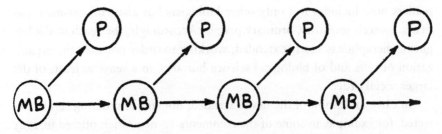

FIGURE 4.6   P = people; MB = molecular biologists.

remarks on eugenics are conspicuously marked by the time in which they were made, and I doubt that he would repeat them today. So much has changed in these last three decades—in ways that go well beyond changing notions of political correctness—as to lend his comments an innocence that now seems quaint. They describe a social order in which the capacity of molecular biologists to "read" the genetic text lends power by virtue of the ability to communicate ("to get across to people"), and by virtue of the authority thus provided to "apply social pressure" (again through speech). Today, however, we have witnessed the realization of a form of power far more direct than any provided by mere acts of speech. Out of the rhetoric of reading and writing genetic texts has emerged a set of technologies for materially altering the resident DNA of living organisms, that is, for actually manipulating genetic "texts." To be sure, such manipulations remain a far cry from any possibility of "rewriting" a genetic text to order—indeed, the effects produced are often so far from what had been anticipated that they seem to confound any such hope for the future. Still, the basic fact is indisputable: with the technologies that became available, knowledge of DNA sequences could be directly and reliably channeled to the production of concrete effects of a sort that an earlier rhetoric of "rewriting" had only envisioned. And this fact in itself has transformed the entire structure of the reading paradigm.

The project to decode the "book of life" was first proposed to Congress in the late 1980s as an extension and fulfillment of the legacy established by the first generation of molecular biologists. Watson was explicit: he described it as "the opportunity to let my scientific life encompass a path from the double helix to the three billion steps of the human genome."[16] Of course, Watson's path from the double helix to the human genome had a long prehistory, indeed as long as that of the *book of life* metaphor itself. And the scientific advocates of this project made good use of that history. The Human Genome Project (HGP) was presented as the natural and inevitable culmination of the age-old quest for knowledge and understanding, and it was sold to the American public (as well as to much of the biological community) under the equally venerable premise (and promise) that such knowledge would inevitably yield power. Specifically, it would yield the power to end human suffering. Further undergirding this promise was the well-established contract between science and society so starkly reflected in the earlier comments of Crick's quoted earlier. This contract came with an implicit guarantee—namely, that the power thus promised, just by virtue of being vested in the hands of disinterested molecular biologists, could be relied on to work for the public good. But by the late 1980s when the HGP was first proposed, the fraying of this hallowed contract was already conspicuous.

By this time, the presence of new players—brought in by the successes molecular biology had demonstrated in delivering tangible goods, and bringing with them new possibilities for reading and writing—could hardly be overlooked, and of course it was not. The importance of these new prospects in selling the HGP is well attested to in the Congressional Record. The phenomenal growth of the biotech industry that followed the development of techniques of recombinant DNA in the 1970s is well documented, and it does not need recounting here. Less familiar, however, and less well documented, are the ways in which the entrance of venture capital has transformed the very nature of biological science. The HGP did not drive these changes, but it surely helped to accelerate them. Perhaps more importantly, it also embodied them. In the evolution of this project we can clearly observe the devolution of the last illusions of scientific control.[17] If for no other reason, Crick would not repeat his comments today because in the world we presently inhabit, it is no longer evident that they make any sense. Who, today, are Crick's "we"? And who are his "they"? Crick's innocent division of the world—into a "we" of scientists and a "they" that includes everyone else—has been dramatically subverted by the strong arm of the market, and in ways that we can all clearly recognize. As molecular biologists read sequences, investors read stock market indexes, and they quickly parlay this information into the targeted investments on which biological research has become increasingly dependent. Sometimes, the reader of sequences and the

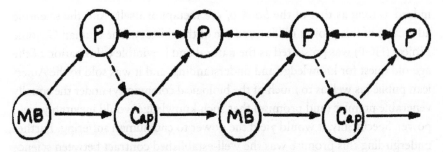

FIGURE 4.7   MB = molecular biologists; P = people; Cap = capital.

reader of stock market indexes may be one and the same person, but it matters little. What matters far more is that, in this process, the authority of the market has come to at least rival the authority of the molecular biologist qua scientist, calling more and more of the shots in real-time decisions about what kinds of genetic texts are worth "reading" and what forms of "rewriting" are worth pursuing. From the vantage point of this chapter, we might represent the new configuration as in Figure 4.7.

To bring my mapping of readers and writers to the present, I want to close with one last ironic twist. Just as the older reading paradigm relating human authors of biological treatises to their readers has been disrupted by the real politics of science in action, so too has the reading metaphor relating genes to cells been disrupted by our achievements in probing the material dynamics of genes in action. Genes no longer simply transmit their messages; in order to function appropriately, they must also be competent to read the many signals transmitted to them by protein and RNA molecules, carrying necessary "information" about the current state of the cell and its neighbors. Similarly, messenger RNA can no longer be thought of as carriers of messages from the DNA that need only to be translated into amino acid sequences, for we now see that, in higher organisms, these very messages require extensive "rewriting" in the course of development—a rewriting performed by protein assemblages specifically designed for the purpose. As Griffiths and Gray observe, "It makes no more and no less sense to say that the other resources 'read off' what is 'written' in the genes than that the genes read off what is written in the other resources."[18]

More radically yet, when the cell is under stress, and normal processes of replication cannot proceed, yet other proteins are called into action to jettison parts of the original DNA sequence and replace it with any sequence at all that will permit replication to proceed. In contemporary molecular biology, the cell has become a vast internet, relaying messages back and forth between and among a dizzying assembly of readers and writers, each of them constituted of complex associations of DNA, RNA, and proteins. To be sure, the DNA sequence must remain intact if we are to have intergenerational fidelity, but guaranteeing the stability of DNA has turned out to be no simple matter. Where fidelity of replication was once seen as a straightforward effect of DNA structure, it is now seen as a consequence of a highly coordinated system of enzymatic reactions that involves the participation of hundreds of proteins.

Is there a moral to this story? I think there is, and it is about the relation between language and action. Let us start by asking, what is a metaphor? For the purposes of my argument, I think I can bypass the sophisticated discussions of metaphor in recent years (but you will surely tell me if I am wrong), and rely instead on an earlier, and simpler understanding in which a metaphor is in the first instance a vehicle, carrying meaning from one referent to another, reflecting one aspect of reality through the lens of another, and back again. As such, it also has a kind of force, forging new ways of perceiving the world, and hence, new ways of acting in and upon the objects of our perception. Similarly, because a metaphor is a way of talking or writing, and itself a communicative act, it is also a vehicle for transmitting new ways of perceiving and acting, new orderings of the world, from an author (or speaker) to his or her audience. But where do metaphors come from? In part they come from the experience of authors as readers of other texts, but they also reflect the experience of authors as actors in a material (nonlinguistic) world. Thus metaphors, even though in language, are at the same time also vehicles for carrying the effects of our experience as material (nonlinguistic) actors into language, and back again. It is just this interactive process, between language and action, that I hope my tracking of the linguistic metaphor in genetics has illustrated. If there is a particular moral to this story, a particular message I am trying to convey, it is this: whether or not empowered as authors, human readers are actors in and on the world; as such, they can and do transform the meaning of the language they read. Apparently, this is also true for the metaphoric readers we find in cells. I am arguing that, just because reading (or listening) cannot be separated from other ways of acting in the world, there is no such thing as a passive reader (in the literal sense of the word) who, however dis-empowered, does not have some way of answering back. The same, it seems, also applies to the metaphoric readers found in the cell—to that vast repertoire of cellular proteins whose existence may depend on a reading of DNA, but whose activity determines when and how a sequence becomes a message, and even the very meaning of the message to be read.

CHAPTER FIVE

The Decline and Fall of Astrology as a Symbolic Language System

Ann Geneva

The debate on language continues to rage; even definitions of the word itself spawn endless polemics. The seventeenth century, however, had no such difficulty. They knew that Latin had been the lingua franca for intellectual, and particularly technical, scientific, and philosophical discussion for centuries.

With Puritan and republican currents convulsing England, by the century's end the use of the vernacular was beginning to be seen as paramount to national identity, not least by the fledgling Royal Society. How best then to remain part of the international community?

The quest began, as it did in other countries, for a universal symbolic language to replace Latin, for a pure language without papist or political undertow. Everyone knew what this meant: symbols would be used rather than words, which had begun to acquire the taint of literary metaphor with its ambivalence, inimical to the newly emerging sciences that were edging their way toward the intellectual center of the world stage.

Thus when I use the term *symbolic language*, I mean simply this: a language that signifies using symbols rather than conventional letters, what the searchers after a universal language would have meant by the term—namely, a nonorthographic system of signs. In this quest, astrology with its claimed monopoly on reducing the universe to symbols, became one of the competing paradigms.

For the astrologers, of course, the stars themselves were the symbols that comprised the language of God, who communicated with his sublunar creatures through their mediation and translation.

I have discussed how astrology functioned as a symbolic language in chapter 6 of my book *Astrology and the Seventeenth Century Mind*, which decodes the starry language courtesy of the century's leading astrologer. This discovery led to some important implications about the way language itself functioned during this time of ferment when scientific language was being formulated. To the virtuosi of the time, the concept was clear, even if the methodology was not. I now leave it to the theorists among you to pick over the bones of this meaty feast of science and language.

While researching English civil war propaganda among the astrological manuscripts in the Bodleian library collected by Elias Ashmole, I found the astrological material beginning to speak of its own accord—in a language encoded in tongues. My research led me ultimately to argue that astrology is best understood, and its decline accounted for most persuasively, by viewing it as a symbolic language system functioning within a diminished though not yet discredited Neoplatonic framework. In doing this study I found myself confronted with not only a lost symbolic language, but an entire universal explanatory system from which the twentieth century had become dislocated.

In an attempt to avoid viewing astrology as only a fringe preoccupation, it may help to borrow a paradigm from a discipline that studies universal explanatory systems. Speaking of worldviews and religions as products of human consciousness, writers in the sociology of knowledge such as Peter Berger and Thomas Luckmann furnish the idea of a humanly conceived symbolic universe, which then presents itself as a full-blown and inevitable entity. Such a worldview is able to integrate different provinces of meaning and encompass the institutional order in a symbolic totality. It thus becomes the matrix for the entire historic society, as well as for the biography of the individual.[1]

For many in seventeenth-century English society, astrology performed just such a function. When employed as it was by an astrologer such as the civil war parliamentarian propagandist William Lilly to interpret the entire design of life to a society, it became a powerful tool for allowing individuals to incorporate belief and experience into the surrounding social, political, and religious nexus. Lilly's casebooks and writings show him pronouncing on questions ranging from pregnancy, battles, safety of ships, and the theft of linen, to the fates of the Archbishop of Canterbury, the King, and the Commonwealth. Their basis in ancient tradition gave these divinations quasi-sacred authority, and their grounding in astronomy and mathematics gave them semi-scientific ratification as well. Everything from the disintegration of the social order to the theft of six silver spoons could be analyzed using the same methodology within a unified framework.

As Franz Boll has noted, after Ptolemy's systematization it was difficult for people to resist "the powerful fascination of such a well-organized system."[2] Lynn Thorndike and C. H. Josten have both argued that before Newton postulated his law of universal gravitation, astrology had been the only generally recognized universal natural law.[3] Lilly himself cited Albertus Magnus to argue, "No humane Science or Learning doth perfectly attain the Ordination of the whole Universe; except the knowledge of the judgement of the Starres."[4]

Such a symbolic universe needs a language with which to interpret its signs in order to create a canon of belief. Astrologers in the seventeenth century employed a system that combined mathematical symbols with astronomical planetary glyphs, integrating it within political and providentialist commentary. Their use of an ancient system of symbols, unique to their art, lent an aura of learned supernaturalism to their claims. God's will was writ large in the heavens, and astrology provided the interpretative methodology. Thus astrology functioned as both a symbolic language and universal explanatory system.

Crucial to an exploration of astrology's position in the context of seventeenth-century language is the distinction between the metaphorical use of astrology—mainly as a rhetorical literary device—and its function as a genuine symbolic language. In the seventeenth century, these two ways of employing astrology (with many gradations) could occur in works by the same writer with little apparent confusion between the two modes. By language system I mean simply a system of interconnecting symbols that signify.

The parliamentarian astrologer William Lilly used astrology's symbolic language as coded substitute for politically dangerous pronouncements. When wishing, for example, to speak of the King without incurring royal wrath, he referred to the tenth house—the traditional dwelling of rulers in astrological figures. In doing so, he drew on long-established astrological precedent while introducing innovations specific to circumstances engendered by the civil war.[5] Before we attempt a discussion, it is important to take a brief look at examples of celestial metaphor versus the actual "starry language" itself. The two might reinforce one another, but they were never identical.

Major elements common to seventeenth-century astrology were found even in metaphorical usages—the sun as signifier of the King; the sign Gemini for the City of London; recognition of the ascendant as the crucial element in an astrological scheme, and so forth. This survey of metaphor is not intended as comprehensive, but highlights some contemporary instances outside the literary mainstream (in which it can sometimes appear difficult to find verses not grounded in astral metaphor). Some of what is often regarded as fanciful analogy actually contains literal references to contemporary astrology. Yet even its metaphorical use, as in Donne's Easter sermon of 1623, indicates astrology's pervasive presence in seventeenth-century language and consciousness.

First, a few disparate examples. Typical of literary wordplay is a passage from John Dryden's *An Evening's Love, or the Mock-Astrologer* first performed in 1668. Like Chaucer in the Canterbury tales, Dryden uses technical astrological terminology to mock the subject in other scenes, but here Wildbloud speaks merely in astral metaphor to Jacintta of his love: "I am half afraid your Spanish Planet, and my English one have been acquainted, and have found out some by-room or other in the 12 Houses."[6] John Milton encapsulated the civil war in more sophisticated astrological metaphor:

> Among the constellations war were sprung,
> Two planets rushing from aspect malign
> Of fiercest opposition in mid sky,
> Should combat, and their jarring spheres confound.[7]

A nice instance of metaphor without sacrificing technical accuracy is found in another astrological battle. In his brief life of Thomas Hobbes, John Aubrey explains his unwarranted exclusion from the Royal Society: "He would long since have been ascribed a member there, but for the sake of one or two persons, whom he tooke to be his enemies: viz. Dr. Wallis (surely their Mercuries are in opposition) and Mr. Boyle."[8] Since Mercury metaphorically represented the mind, this perfectly denotes Hobbes's and Wallis's mental opposition; and since Aubrey cast nativities for as many of his biographical subjects as possible, it may well be that their planetary Mercuries were physically opposed to one another as well.

A sermon preached by Archbishop William Laud for the opening of Parliament on 17 March 1628, concluded with the following lines:

> Ioyne then and keepe the Unity of the Spirit,
> and I'le feare no danger though Mars were
> Lord of the Ascendent, in the very instant of
> this Session of Parliament, and in the second
> house, or ioyned, or in aspect with the Lord
> of the Second, which yet Ptolomey thought
> brought much hurt to Common-wealths[.][9]

While these figures of speech certainly enhance Laud's rhetoric, their technical nature indicates an awareness of astrological discourse, both on his part and presumed on that of his audience. Laud reveals himself well enough acquainted with the niceties of astrology to know the baleful influence Mars would have as Lord of the ascendant in such a scheme. This passage also shows Laud's familiarity with the practice (adhered to by Lilly and other political astrologers) of striking a figure for the opening of Parliament in order to ascertain its potential success or failure.

A speech by Clarendon in Parliament in September 1660 followed the King's and was meant to assure Parliament that all was now well among the constellations. In contrast to Laud, however, Clarendon's astrological references point less to his own familiarity with the art than to an awareness that astrologers—including those formerly scanning the heavens for the parliamentary cause—were busy positioning the Restoration within an astral design. Here he is clearly playing on both astrology's waning influence and its still common currency to express hopes that would have gladdened Laud as well:

> The Astrologers have made us a fair Excuse; and, truly, I hope, a true one:
> All the Motions of these last Twenty Years have been unnatural, and have
> proceeded from the evil Influence of a malignant Star; and let us not too much
> despise the Influence of the Stars. And the same Astrologers assure us, that the
> Malignity of that Star is expired; the good Genius of this Kingdom is become
> superior, and hath mastered that Malignity; and our own good old Stars
> govern us again: And their Influence is so strong, that with your help they
> will repair in a Year, what hath been decaying in Twenty.[10]

Indeed at that time all the astrologers were heralding Charles II's return with the same metaphysical certainty as his father's death had inspired. But Clarendon's intention was more to provide a rhetorical vehicle for the declaration of amnesty.

The equation of the sun as metaphor for a king was always effective. James Howell in 1644 used solar symbology to attribute Charles's eclipse to his having misdirected his beams.[11] At the end the author offers an explanatory key: "Moral. Such as the Sunne is in the Firmament, a Monarch is in his Kingdome:". Similarly, an anonymous royalist pamphlet of 1642, "The English Fortune-Teller," capitalized on public knowledge of Gemini's rulership of London: "The hopefull twins, the upper and lower House (as legs) are the best props and pillars of our crackt State, so long as the Signe is in Gemini."[12]

It also spoofed eclipse predictions: "There will happen a great Eclipse of our Moon-fac'd changing Common-wealth, visible in our Horizon, by the interposition of ill Counsellors, who strive to obfuscate the lights of reason and kingdome. The whole time of duration is till there be an union."[13]

And so forth and so on.

Despite their power to exalt the rhetoric, none of these uses, however technical, compares with Lilly's manipulation of the symbols of astrology to constitute a genuine language system. Having looked at a series of metaphors grounded in the practice of astrology, let us examine briefly some cases of nonmetaphorical astrological syntax. That the heavens possessed their own peculiar grammar, embodying a pattern of intelligible symbols meant to be decoded by human cryptanalysts, underlies much seventeenth-century thought. Probably unknown to Lilly, there was even a tradition of celestial alphabets where the forms of the letters were derived from observation of configurations of stars in the heavens which could be "read" as a form of sacred writing.[14] John Bainbridge based his interpretation of the comet of 1618 upon what he termed "celestial hieroglyphics." Although Bainbridge took care to distinguish between these and "vulgar astrology," the term he employed is the most accurate and fruitful perspective from which to view seventeenth-century astrology, which Keith Thomas noted was often linked to the companion study of cryptography.[15]

The basis for the astrologers' authority was the premise that if the Creator wished to communicate with his creation, what better method could be devised than employing the apex of his creation—the heavens—as a vast celestial billboard. The astrologers' role was to decode these enciphered messages by translating the celestial paradigm into human language. Sermons were preached in support of this basis for astrology by divines who pointed out that when God wished to send his son to earth to redeem humankind, he used a blazing star to indicate the divine birth.

Astrology's diagnostic system of symbols was regarded as the key to the organizational harmonies of the universe. If properly interpreted its symbols could be decoded and translated into words that would reveal God's plans for his far-flung sublunar creatures. It was this decoding ability that constituted the adept astrologer's genius. Astrology as a language system occupied the middle ground between the abstract symbolic system of classical mathematics and common discourse, as it did between divine and human language. The civil war astrologers were able to adapt astrology's symbolic system to political paradigms that they then interpreted and translated into sometimes singing vernacular prose.

In a lecture on the concept of polarity in the life and thought of early modern England, Keith Thomas exhorted historians to discover the "mental grammar" of earlier periods.[16] It is this one must come to terms with in the case of seventeenth-century astrology, although that century did not invent the notion of astrology as language. Plotinus was highly critical of astrological practice in his day. Yet he offered this explanation of heavenly discourse within the ordered Neoplatonic universe on which astrology was metaphysically dependent:

> But if these planets give signs of things to come—as we maintain that many other things do—what might the cause be? How does the order work? There would be no signifying if particular things did not happen according to some order. Let us suppose that the stars are like characters always being written on the heavens, or written once for all and moving as they perform their task, a different one: and let us assume that their significance results from this. . . .
> All things are filled full of signs, and it is a wise man who can learn about one thing from another.[17]

This commentary constitutes a clear example of the worldview that provided the grounding for astrology as it developed after Ptolemy: that of a perfectly ordered universe in which each element reflected and signified all others. Its organization was not the taxonomic austerity of modern science. Rather it was structured by the Pythagorean notion of musical harmonies, the medieval doctrine of signs, and the idea of a natural hierarchy of creation with the heavens at the apex. Astrology's universe was not circumscribed by the triumvirate of astronomy, mathematics, and physics as is commonly held.

But it is the equation of the "stars" (that is, planets and stars) with written characters that claims our attention. Astrologers themselves were conscious they were using a language. The astrologer George Atwell's *Apology* unwittingly echoed Plotinus: "These Phaenomena's are Gods Embassadors to a land, and that land is in an ill case that hath never a one that knowes their language."[18]

William Lilly commented on "the ascendant at tyme of the lesser Conjunction of ♄ and ♂ in May this yeare: by all w^ch, wee are tould in the heavenly language, how great the Actions and how terrible the warrs shall bee of this yeare."[19] Another time he focused on English stars: "But let us now close home to our English Parliament, and tell the Grandees thereof, the language of heaven, and the portent of this present Appearance, in as plaine and familiar a Dialect of truth, as we may."[20] Again he stated in an almanac, "The English Nation (if our Starry language fail us not) hath several reports from the North east."[21]

In another pamphlet, Lilly wielded the language of astrology like a pikestaff. Commenting on the eclipse in August 1645 he underscored his encoded use of astrological language in simultaneous translation; speaking covertly of London through its Gemini ascendant, Lilly proclaimed, "Countrey-men, I tell you, Gemini is in the eighth House of this Eclipticall Figure; but Mercury, that is Lord thereof, is in Virgo, his own house, angular, and departing from the Sun-beams. Wilt thou have me speak plain English? This very Citie must suffer a share in the effects of the Eclipse, it were a wonder else, the Times considered."[22]

This tradition of encoded prophecy reflected a Neoplatonist perception of the universe as fundamentally interconnected, but enciphered rather than transparent as the emerging sciences conceived it. Indeed, despite Lilly's didactic efforts to make it familiar to all, the astrological system remained best suited to concealment. By the century's end, metaphysical revelation became better served by the developing tradition of empirical science than it had been by religion or astrology.

In this new world, astrological symbology suffered from its multiple denotations. Interpretively, the symbol for Mars, depending on its context, could in seventeenth-century astrology signify (among other referents) the head, the color red, anger, war, butchers, hangmen, the metal iron, wounds, the element fire, the chestnut tree, the shark, the raven, west winds, scars, choler, fever, the left ear—and, of course, death. The King's symbolic representation could be not only the sun, but Saturn, Jupiter, the tenth house or the rulers of any of these, as well as being signified by individual elements of his own nativity.[23]

All these were legitimate designations within the astrological universe of discourse, embedded as it was in a Neoplatonic interrelated plenum. Our study of astrological encoding points the way to a solution for astrology's inevitable decline. The very wealth of signification that enabled astrologers to select from among multiple designations caused the starry language to be perceived as ambiguous and enigmatical, especially by its detractors, when exposed to the light of day. It also legislated astrology's defeat amid the competing linguistic paradigms of mid-seventeenth-century England.

Astrology was logical, if complex, and consistent, if multilayered—but transparent it was not. Satirists railed at the language of this art whereby one "should not finde one answer but did include a twofold meaning"[24]: "Your discourse . . . hath no Realitie or Essence in it: but you huddle a companie of Astronomicall words together, wanting both Coherence, Methode, and Congruitie; you powre out whole Dictionaries of strange Words, talke as though you could repeat. . . . English Hollingshead without booke."[25] The astrologer is also accused of speaking in tongues: "The Mountebanks Drug Tongue, the Souldiers bumbasted Tongue, the Gypsies Canting Tongue, the Lawyers French Tongue, the Welch Tongue; nay, all the Tongues that were at the fall of Babylon (when they were all confusedly mingled together) could as well be vnderstood as [t]his strange Tongue."[26]

It has been argued that "the fundamental impulse of the [seventeenth] century was towards the 'explanation' of what had hitherto been mysterious; towards the statement in conceptual language of what had hitherto been expressed, or imagined in pictures and symbols."[27] Astrology's symbolic language could not be arranged within conceptual categories. Its meaning could not be reduced to physical components, making it impervious to quantifiable analysis. Its signs and symbols, locked into meanings by ancient decree and cultural transmissions, and dependent on its own signifying system, were not susceptible to the new empirical experimental methods or proofs.

Astrology's defeat, then, can be viewed in two dimensions. Its failure as a universal explanatory system was accelerated by the dismantling of a Neoplatonic universe honeycombed with inextricable consonant harmonies. As a symbolic language system its multiple denotations reflected the universe it signified, each sign's meaning mirroring an animistic hierarchy of being.

The cross-rip that developed between an innate impulse to conceal and a demystifying effort to reveal ultimately engulfed astrology. This conflict was echoed in the ferment of mid-seventeenth-century England concerning the nature of language, and specifically in the quest for the development of a universal character and philosophical language. Astrologers were aware of translating from the celestial language of the Creator into human discourse. The relationship between these linguistic spheres was central to debates on the nature of language in the seventeenth century. These debates can be taken as a context for our discussion of astrology's demise as an elite symbolic language system, a failure that brought down the entire Neoplatonic edifice like a house of fortune-teller's cards.

There is an entire body of literature on the topic of universal language movements in the seventeenth century, but we can profit by examining some norms inherent in the attitudes of those concerned with the invention of a universal language, attitudes that had irreversible consequences for the fate of the astrological paradigm.

The basis of the movement was Francis Bacon's dictum against the Idols of the Marketplace. Since words—which had come into existence through the "apprehension of the vulgar"[28]—did not accurately mirror reality, language was one of the main obstacles in the communication of knowledge. Bacon was certain that a universal language system could be devised in which characters would be assigned to signify individual entities. Unsurprisingly, Bacon looked to the mathematical sciences as his model. Bacon praised the mathematicians' use of definitions, but since even they entailed a use of words, he gravitated in *De Augmentis Scientiarum* to the use of symbols that signify things directly.[29]

The first work to speak of a universal character was Wilkins's 1641 *Mercury, or the Secret and Swift Messenger*. Ironically, one discipline he mentioned that used universally understood signs was astronomy, whose tradition he connected to an older secret astrological one:

> The Astronomers of severall Countries doe expresse both the heavenly Signes, & Planets, & Aspects by the same kind of notes. As, ♈, ♉, ♊, ♋, &c. ♄, ♃, ♂, ♀, &c. ☌, ✳, △, □, ☍. Which characters (as it is thought) were first invented by the ancient Astrologers for the secrecie of them, the better to conceale their sacred and mysterious profession from vulgar capacitie.[30]

Wilkins then explained that although different countries would have differing words for each sign, the sense would be identical.

The universal language movement can be seen as spanning publication of two works by Wilkins—from his *Mercury* in 1641 to *An Essay Towards a Real Character And a Philosophical Language* in 1668, the magnum opus of the language movement. After its publication the Royal Society formed a committee to establish a developmental framework based on its ideas. Committee members included John Wallis, John Ray, Robert Boyle, Christopher Wren, Seth Ward, Robert Hooke, and the diarist John Evelyn; Wallis and Hooke actually used the character put forth in the *Essay* for scientific purposes.

In 1641 Wilkins was already claiming that existing language was unsuited to scientific purposes because of ambiguities and irregularities. He advocated creation of a universal language that would accurately reflect nature and at the same time provide a medium for scientific communication.[31] In 1652, Francis Lodwick proposed a new form of language in *Of an univer-sall reall caracter*, the first attempt to create a genuinely scientific language. Such a language, he claimed, would be of great assistance to natural philos-ophy, but—tellingly—not to poetry.[32]

This distancing of natural philosophy from poetry reinforces our ultimate conclusion that astrology's destiny is related to poetry and metaphor. Even the poet Cowley, although writing in ornate prose, did not remain unaf-fected by Bacon's elevation of things over words as seen in this poem prais-ing him, "To the Royal Society," printed in the front of Sprat's *History*:

> From Words, which are but Pictures of the Thought,
> (Though we our Thoughts from them perversly drew)
> To things, the Minds right Object, he it brought,
> Like foolish Birds to painted Grapes we flew;
> He sought and gather'd for our use the Tru[.][33]

Also taking his cue from Bacon's dissatisfaction with verbal ambiguity, George Dalgarno, a Scottish schoolteacher living in Oxford, expressed his wish to invent a language "free from all anomaly, aequivocation, redun-dancy and unnecessary Grammatications: and the whole institution being suited to the nature of things."[34] This reveals the developing split between ordinary and scientific language, eventually so lamented in Pope's *Dunciad*.

The major stumbling block of the universal language movement was the presumed isomorphism between words and what they signified. Seth Ward viewed words as troublesome obstacles, to be overcome not by rediscover-ing an original self-signifying language, but by creating syntactical taxon-omy. Certain that "Symboles might be found for every thing and notion," he was aware that the number of characters needed would be "almost infinite." The solution to his dilemma was to compound the symbols and arrange them in a hierarchy following the principles of mathematics: "This design if perfected, would be of very great concernement to the advancement of Learning, and I know one in this University [Oxford], who hath attempted some thing this way, & undertak[e]s as farre as the tradition of reall Learn-ing, by which I understand the Mathematicks, and Naturall Philosophy, and the grounds of Physick."[35]

Pitted against these Baconians was another faction of the universal language movement with diametrically opposite aims. Their attempt to reinfuse language with its original sacral dimension seems to have been one consequence of the mid-seventeenth-century Neoplatonic revival. The arcane verbal and pictorial creations of Renaissance alchemists have been seen to "represent a sustained effort to evoke an experience of a kind for which the syntactical structure of conventional language is unsuited."[36] A similar view has been taken of the Cambridge Neoplatonists' attempt, fearing the life of the spirit was vanishing, to revive what was vital in the religious tradition: "For this purpose it was convenient to change the linguistic currency: to speak of religious matters in terms other than those in constant use, and . . . to think of them in modes whose very possibility . . . showed up much contemporary thought as inadequate or crude."[37]

John Webster's *Academiarum Examen* argued against an emphasis even on biblical tongues at the expense of the true divine language—the enciphered discourse of the Holy Spirit: "So that he that is most expert, and exquisite in the Greek and Oriental tongues, to him notwithstanding the language of the holy Ghost, hid in the letter of the Scriptures, is but as the Hiroglyphicks, and Cryptography, which he can never uncypher, unless God bring his own key, and teach him how to use it."[38] F. M. van Helmont, son of the famous iatrochemist Jan Baptista van Helmont, wrote a utopian work, *Alphabet of Nature*. This work encapsulated the essence of that part of the universal language movement that believed that symbols could authentically reflect nature. This faction, of which Webster was a conspicuous representative, grew from the ideas of Hartlib, whose Invisible College counted Petty and Boyle among its members, as well as Comenius. It sought in its formation of a language system to uncover the inherent, self-identifying divine signatures of the creation.

This was in turn a version of the Renaissance Christianization of an animistic tradition, most clearly expressed in the Paracelsian idea of the universe as a vast alchemical unfolding. According to Paracelsus each created thing signaled its inner essence, and the true "physician" through purity of spirit was able to recognize and decode these messages. The original Adamic language had captured and expressed these prelapsarian signatures in its letters and names. This notion of an inherent sacred language was one—the unprevailing one—of the two major factions within the seventeenth-century quest for a universal character and philosophical language to replace Latin as the lingua franca of the intellectual elite.

Webster, a chemist in the Paracelsian tradition, viewed the quest for a universal language as an attempt to recapture the original "Paradisical language"—known to Adam but lost at the Fall—which was capable of being reflected in an earthly one:

> There are, it may be so many kinds of voices in the world, and none of them mute, or without signification. Many do superficially and by way of Analogy (as they term it) acknowledge the Macrocosm to be the great unsealed book of God, and every creature as a Capital letter or character, and all put together make up that one word or sentence of his immense wisdom, glory and power.

Webster proposed the discovery of a "universal character" by means of a return to the tradition of hieroglyphics and cryptograms, embodied in the ancient wisdom of the *prisca theologia*. Webster judged that universally understood symbols such as those for shorthand, astronomical, and mathematical symbols indicated the way to transcend the limitations of individual languages. He also noted stenographic and algebraic symbols, Chinese ideographic characters, deaf language, and so forth, arguing that they should be used as models in the realization of a universal character: "If we do but [c]onsider that the numeral notes, which we call figures and cyphers, the Planetary characters, the marks for numerals, and many other things in Chymistry, though they be alwaies the same and vary not, yet are understood by all nations in Europe, and when they are read, every one pronounces them in their own Countreys language and Dialect."[39]

The original Adamic linguistic unity could be revived only through decoding the signs of creation. Webster rhapsodized about the "right reading of these starry characters," lamenting current ignorance of celestial language:

> Alas! we all study, and read too much upon the dead paper idolls of creaturely-invented letters, but do not, nor cannot read the legible characters that are onely written and impressed by the finger of the Almighty. . . . And yet indeed they ever remain legible and indelible letters speaking and sounding forth his glory, wisdome and power, and all the mysteries of their own secret and internal vertues and qualities, and are not as mute statues, but as living and speaking pictures, not as dead letters, but as preaching symbols.[40]

Elias Ashmole is sometimes credited as the first to formulate the concept of a universal language. Valuing both the practical and occult qualities of symbols, it is no coincidence that he was also an astrologer. In a 1650 alchemical tract Ashmole recommended that in order to reconstitute the original linguistic unity, "we look back upon those steps already laid to our hands; for we may draw some helps from the Egyptian Hieroglyphicks, Symbols, Musical Notes, Stenography, Algebra, &c. Besides, we see there are certain Characters for the Planets, Signes, Aspects, Metals, Minerals, Weights, &c. all which have the power of Letters, and run currant in the understanding of every Language."[41]

Webster and Ashmole, then, like Dee and Hartlib, believed that the end of natural philosophy was not merely to acquire knowledge of accidental or even essential properties of things. Its purpose as well was to divine what these things signified beyond themselves. This knowledge would ultimately lead the adept to the language of "him, who hath set all these things as so many significant and lively characters, or Hieroglyphicks of his invisible power, providence, and divine w[i]sdome, so legible, that those which will not read them, and him by them, are without excuses."[42]

The alternative view of language—the one which prevailed—had a differing view of the nature its language was to reflect. This quest grew out of "the need for a universal means of communication necessary both as nomenclature for new discoveries about the natural world and as a medium for dissemination of scientific information."[43] Based on scientific and taxonomic concerns, it attracted natural philosophers such as Robert Boyle, William Petty, and John Ray to the search. There was a direct correlation between the growing scientific hegemony and its dominance of linguistic development.

M. M. Slaughter demonstrates the grounding of the universal language movement in biological taxonomy and predicates a basic connection between empirical science and this quest for a new language. "Science," she says, "provides a view which implies that language reflects properties of a real world rather than of the mind."[44] She argues that "for the most part the motivation of the language projectors was more scientific than linguistic; their concern was more with nature than with language."[45] From the beginning, the search for a universal language was connected to the need for an appropriate scientific medium.

The 1650s in Oxford saw great activity in language planning among many who were to form the nucleus for the Royal Society during the next decade. One of their first works was a rejoinder to Webster. Written by Seth Ward, Savilian professor of astronomy, with the aid of John Wilkins, already famous as a Copernican, *Vindiciae Academiarum* engendered what has become known as the Ward-Webster debate. Tone-deaf to Webster's Neoplatonic advocacy,

Ward penned an impassioned defense of university curricula, dismissing Webster's ideas as fantastical. Wilkins's introduction established the tenor:

> What a loose and wild kind of vapouring is that C[h]ap. 3 about Cryptography, and the universall Character. . . . But above all . . . that canting Discourse about the language of nature, wherein he doth assent unto the highly illuminated fraternity of the Rosycrucians In his large encomiums upon Jacob Behem, in that reverence which he professes to judiciall Astrologie, which may sufficiently convince what a kind of credulous fanatick Reformer he is like to prove.[46]

Ward continued this line of argumentation against Webster "who taking of Hieroglyphicks, Emblemes, Symbols, and Cryptography, and according to his capacity, hath extracted out of silence, an advance Eloquence, and from dumb signes of Grammar."[47] No sermons from stones for Ward. He further mocked Webster's notion "that Tongues, nay letters, have taught a way of Mysticall Theology."[48]

Ward then zeroed in on Webster's central transgression, catching him in the collision between the impulses to encrypt and to reveal:

> Hierogliphicks and Cryptography, were invented for concealement of things, and used either in mysteries of Religion which were infanda, or in the exigenc[i]es [o]f Warre, or in occasions of the deepest Secresy . . . and Grammar is one of those Arts and Language one of those helps, which serve for explication of our minds and notions: How incongrous then is it, that the Art of Concealment, should not be made a part of the Art of Illustration.[49]

Ward here denies any arcane dimension to grammar and language, also spotlighting the fundamental contradiction in Webster's attempt to reveal through cryptograms.

Thomas Sprat in his 1667 *History of the Royal-Society* gave his opinion of the tradition of secrecy and initiation of the Assyrians, Chaldeans, and Egyptians:

> It was the custom of their Wise men, to wrap up their Observations on Nature, and the Manners of Men, in the dark Shadows of Hieroglyphicks; and to conceal them, as sacred Mysteries, from the apprehension of the vulgar. This was a sure way to beget a Reverence in the Peoples Hearts towards themselves: but not to advance the true Philosophy of Nature. That stands not in need of such Artifices to uphold its credit.[50]

Sprat thus felt that natural philosophy should brook no mediation. He wanted the knowledge of nature demystified, specifically pitting the new philosophy against the astrological mentality:

> All Knowledg is to be got the same way that a Language is, by Industry, Use, and Observation. It must be receiv'd, before it can be drawn forth. 'Tis true, the mind of Man is a Glass, which is able to represent to it self, all the Works of Nature: But it can onely shew those Figures, which have been brought before it: It is no Magical Glass, like that with which Astrologers use to deceive the Ignorant.[51]

Views such as those expressed by Ward and Sprat were not based entirely on linguistic objections. As radical dissent had gathered speed from the depths of social discontent and prophetic inspiration, it had found existing vocabularies wanting. Neither traditional theological, historical nor literary language proved adequate to encompass the psychic dislocation of the time. It was into this linguistic void that the excesses of prophetic language generally, and astrology's prognostic absolutism in particular, burgeoned. The result was the extreme expressions of the various sects and movements as seen in the pamphlet explosion between 1642 and 1660. Lilly had combined the inflated language of prophecy with that of astrology, which for all its domesticated preoccupations was still preeminently suited to extremity, including death and providentialism.

The faction of the universal language movement represented by Wilkins, Lodwick, Ward, and Wallis insisted that language was based not in divine hierarchy but on human convention. The criteria they formulated were ultimately given voice in the Royal Society's linguistic norms put forth by Sprat in 1667. By the time his *History of the Royal-Society of London* was published, Anglican preachers were railing from their pulpits against the overblown rhetoric that had been used by the Puritans. Thus linguistic excess spawned by sectarian fanatics was a target, as well as the idea of language as divine signification.

Wilkins's 1668 *Essay*, dedicated to the Royal Society's council and its fellows, condemned the unbridled expression that had been given to claims of the various religious factions. Sprat lamented that the English language, which had begun to raise itself since the time of Henry VIII, had in the time of the recent civil wars "receiv'd many fantastical terms, which were introduced by the Religious Sects; and many outlandish phrases." Now that passions had been allayed, it was time for "some sober and judicious Men" to purify the tongue.[52]

Webster was decidedly not regarded as the kind of sober and judicious man the project required. Unsurprisingly, he also promoted astrology as an art that embodied the highest form of aspiration: "What shall I say of the Science, or art of Astrology . . . ? For the art it self is high, noble, excellent, and useful to all mankind, and is a st[u]dy not unbeseeming the best wits, and greatest scholars, and no way offensive to God or true Religion."[53] It may not be irrelevant that Webster had served as surgeon and chaplain during the civil war to the parliamentary army, for whose cause so many of the astrologers had marshaled their prophecies.

Ward, equally unsurprisingly, did not prove an admirer of astrology, and countered Webster's diagnosis of the universities' failings thus:

> But the mischiefe is, we are not given to Astrology, a sad thing, that men will not forsake the study of Arts and Languages, and give themselves up to this high and Noble Art or Science, he knowes not what to call it: Nay call it that ridiculous cheat, made up of nonsence and contradictions . . . so as none but one initiated in the Academy of Bethlem [Bedlam] would require of us.[54]

Significantly, Ward saw Webster's embracing of astrology (which he lumped with magical arts) as a betrayal of mathematics:

> I have an itching desire to know what Lilly, and Booker, Behmen, and all the families of Magicians, Soothsayers, Canters, and Rosycrucians, have done to vexe him, since he was writing of Mathematicks, and Scholastick Philosophy, that having cherished them . . . he should now of a sudden cast them off, betaking himselfe to their deadly enemy.[55]

Ward thus posited astrology and mathematics as conflicting systems. He took Webster to task for misrepresenting mathematical symbology by pressing it into the service of the Neoplatonist tradition, charging that those like Oughtred "who have exempted themselves from the encombrances of words" little thought "that their designation of quantities by Species, or of the severall waies of managing them by Symbols . . . should ever have met so slight a considerer of them, as should bring them under Grammar." Ward continued:

> It is very well known to the youth of the University, that the avoiding of confusion or perturbatio of the fancy made by words, or preventing the los[s] of sight of the generall reason of things, by the disguise of particular

nu[m]bers, having passed through severall formes of operation, was the end and motive of inventing Mathematicall Symbols; so that it was a designe perfectly intended against Language and its servant Grammar, and that carried so farre, as to oppose the use of numbers themselves, which by the Learned, are stiled Lingua Mathematicorum.[56]

Ward here speaks of numbers being styled as a mathematical language, contrasting mathematical symbology with the confusing babel of quotidian language. The direction in which this science-oriented wing of the universal language movement consistently pointed was away from language polluted by common usage and toward an ideal language, whose characteristics made it sound suspiciously like mathematics itself. For them, language did not involve coaxing out a significance already inherent—it was simply humanly created convention.

Only recently disentangled from the number mysticism of the Agrippan arch-magus, apparent as late as John Dee's *Mathematical Preface*, mathematics seems to have been an unacknowledged norm within the language movement from its inception. An earlier objection to mathematics being taught in universities had been its association with astrology; indeed the two terms—mathematician and astrologer—had often been used synonymously from ancient times. When John Dee's neighbors attacked his house, they destroyed his huge mathematical quadrants with no less fury than his alchemical equipment.

Hartlib referred to a lost universal language scheme, *Wit-Spell*, whose character had been based on mathematical notation.[57] Leibniz dismissed "magical" characters such as those put forward by Dee in his *Monas hieroglyphica* as useless. Leibniz judged ordinary words too imprecise, concluding that only the *notae* of the mathematicians sufficed for accurate representation in a universal language.[58]

Although he had reservations about mathematics, Bacon admired the precision of its language. In the *Advancement of Learning* he stated, "And lastly, let us consider the false appearances that are imposed upon us by words . . . so as it is almost necessary in all controversies and disputations to imitate the wisdom of the Mathematicians."[59] Galileo was convinced that nature could be understood perfectly, if not completely, only through mathematics. The Book of Nature, he stated, "is written in the language of mathematics, and its characters are triangles, circles, and other geometric figures without which it is humanly impossible to understand a single word of it."[60]

Three years before Ward's publication of *Vindiciae Academiarum*, Thomas Hobbes had expressed thoughts similar to those of Bacon and Ward concerning the nature of ordinary words in his Leviathan, also looking toward mathematics for the solution:

> Seeing then that truth consisteth in the right ordering of names in our affirmations, a man that seeketh precise truth, had need to remember what every name he uses stands for; and to place it accordingly; or else he will find himself entangled in words, as a bird in limetwiggs. . . . And therefore in Geometry, (which is the onely Science that it hath pleased God hitherto to bestow on mankind,) men begin at settling the significations of their words; which settling of significations they call definitions.[61]

Hobbes here was speaking of defining terms. When Leibniz turned to mathematics in his search for characters to represent reality, he proposed *characteristica* whereby lists were to be compiled of all essential notions of thought, to which symbols or characters would be assigned. Even religious problems would submit to this notion for a universal calculus. Hobbes hoped that his version of a philosophical language would be able to reconcile religious differences that were based on semantic misunderstandings.

It is in the context of the previous discussion on mathematical denotation that we must view Thomas Sprat's famous dictum for Royal Society language norms. Sprat praised the French Academy for its purification of language, resulting in what Sprat termed masculine, chaste, and unaffected prose. The English genius, he judged, liked best "to have Reason set out in plain, undeceiving expressions."[62] In a section entitled, "Their manner of Discourse," Sprat described members of the Royal Society as distrustful of common language, fearing their intellectual vigor would be sapped "by the luxury and redundance of speech."[63] Sprat saw ornaments of speech "in open defiance against reason" and denounced "the vicious abundance of Phrase, this trick of Metaphors, this volubility of Tongue, which makes so great a noise in the World."

Despairing of linguistic cure in other parts of learning, Sprat still had hope for natural philosophy:

> It will suffice my present purpose, to point out, what has been done by the Royal Society, towards the correcting of its excesses in Natural Philosophy; to which it is, of all others, a most profest enemy.
>
> They have therefore been most rigorous in putting in execution, the only

Remedy, that can be found for this extravagance: and that has been . . . to
return back to the primitive purity, and shortness, when men deliver'd so
many things, almost in an equal number of words. *They have exacted from all
their members, a close, naked, natural way of speaking; positive expressions;
clear senses; a native easiness: bringing all things as near the Mathematical
plainness, as they can.*[64]

Alistair Crombie has argued that "from one point of view, the whole history of European science from the twelfth to the seventeenth century can be
regarded as a gradual penetration of mathematics (combined with experimental method) into fields previously believed to be the exclusive preserve
of 'physics.'"[65] And between the seventeenth and twentieth centuries, modern mathematics has established itself as the universal language of the scientific community.

Ultimately the application of mathematical norms to intellectual discourse
resulted in the virtual disappearance of natural language from the realm of
scientific concourse (although physics at least retained a tradition of verbal
explication through the nineteenth century). Mechanistic physics offered a
new universal law, and notational mathematics became the lingua franca of
the scientific elite, leaving ordinary language to history and belles-lettres.
Slaughter supports this conclusion in her statement that whereas natural history found its proper expression in scientific classification, "the proper representation or expression of mechanistic philosophy is mathematics."[66]

Sprat anticipated this development by his vow "to promote the same
rigid way of Conclusion, in all other Natural Things, which only the Mathematics have hitherto maintained."[67] The mathematician Henry Briggs was
unhappy with the uncertainty of astrological prediction, as was Flamsteed.[68]
The astrologer William Lilly, however, remained untroubled by the discrep-
ancy between astrological prognostication and empirical or mathematical
certainty: "My arguments are not demonstrative, or can be made so: I acknowledge my Prognosticks to be only grounded upon conjectural proba-
bilitie, and are not subject to the senses, or Geometricall demonstration; this
I speak to avoyd carping."[69]

Before mathematics could assume its position as the new high priesthood,
however, the Neoplatonic universe had first to be dismantled. For the mechanistic universe to triumph, it first had to undergo what Slaughter, in a linguistic framework, speaks of as decontextualization. As she puts it, before it
can be reconstituted into a hierarchy, "a thing must first be removed from the

immediacy of its environment, removed from all the contexture, mystery and complexity of life." The thing, she says, is "'decontextualized' in order to be held up as an object of observation and analysis." This taxonomic empirical norm was antithetical to the universe on which astrology depended for its existence, where reality constituted a plenum. This second Fall, which severed microcosm from macrocosm, eliminated astrology as a system of both language and explication.

The characteristics that had rendered astrology so fertile for prognostication and encoding—its richness, flexibility, and capacity for multiple denotation—were the very qualities that legislated its failure as an explanatory paradigm in a century bent on establishing induction, observation, and experimentation as its criteria. And the astrological reformers, including Bacon, who had attempted to bring astrology into line with the new scientific discoveries and criteria met the same fate as the universal language planners: no one ever bothered to carry out their ambitious experiments.

Astrology did not die from satirists' barbs or from internal disagreements within the art. While partaking of the momentum of prophecy, it did not fail as a religion. Despite its grounding in mathematics and astronomy and Baconian rescue attempts, it did not fail as an unfalsifiable science. Since it comprised its own tradition this alone must bear responsibility for its demise, with the cause more satisfactorily ascribed to internal mechanisms. Viewed in this way astrology's failure is that of a symbolic system that had become out of tune with its frame of reference. Its wealth of multiple signification was rendered a liability, reflected in the quest for a transparent, one-to-one, nonmetaphorical language—a language ultimately enunciated by modern mathematics.

Astrology had always provided the translation between divine and human language in an interconnected universe, whose stars and planets occupied the place where prime mover and earthly creation intersected. The failure of astrology was ultimately that of universal metaphor—the loss of mysterious, divinely tuned, and interdependent spheres. When these crystalline spheres fractured into shards, astrology shattered with them.

The seventeenth-century Neoplatonic revival was the last light of a dying star, a doomed attempt to reinvest an already fatally secular cosmos with the sense of sacred home. The perception of divine immanence that had survived the Reformation ultimately could not escape the black hole of mechanistic requirements, which jeopardized even the transcendent Lord of creation. Once empiricism had penetrated the aether, it pointed the way to more powerful technological violation, ultimately revealing—as with the wizard of Oz—not a heavenly host but mere white dwarfs.

Astrology's failure to meet seventeenth-century natural philosophical and linguistic norms, however, did not result in its death but only its disappearance from intellectually elite circles. Lilly's leveling efforts helped ensure its continuation at the hands of ordinary people, where it remains today. But by the end of the seventeenth century, astrology's demise left only poetry to safeguard the metaphorical conception of the universe and to decipher what Robert Penn Warren called "the world's tangled and hieroglyphic beauty."[70]

CHAPTER SIX

Modernity and Instruments Made to Speak a Language That Any Could Understand: The Case of Réaumur's Comparable Thermometers

Christian Licoppe

In 1730 Antoine Ferchaut de Réaumur, then probably the most influential man at the Académie Royale des Sciences in Paris, published a memoir providing "Rules to make thermometers of comparable degrees." To underline the necessity of such rules and such instruments, Réaumur relied strikingly on a language metaphor to describe the practices of temperature measurements. He argued in the following way: "The manner in which they [previous, noncomparable, thermometers] express themselves, so to speak, being all different, one can only understand the language of a thermometer one has followed for several years, and one can in no way understand the language of any other. . . . Not only is one unable to understand the language of other thermometers, but one understands only very little that of his own."[1] The aim of this chapter, or perhaps more aptly, of this historical essay, is to take Réaumur at his word and to consider seriously his repeated, self-conscious ("so to speak") use of the language metaphor, to analyze some of the consequences of Réaumur's work, and explore the interplay between comparable meteorological measurements and travelers' accounting practices as a particular episode in the construction of modernity. I will therefore try to put Réaumur's work on comparable thermometers into a suitable perspective and move on to study the innovative textual use of measurements made with allegedly comparable meteorological instruments by some of the travelers to whom he gave some of his thermometers: the French geodetic expedition in Peru to determine the shape of the Earth, and a few later interesting accounts such as Charles and Gay-Lussac's reports of their balloon ascents at the turn of the century. I will eventually argue that the comparability of thermometers paved the way to a new mode of legitimizing accounts of distant travels and experience, mediated by comparable measurements, which operated quite distinctly from the virtual-witness literary technology that had been both a resource and an accomplishment of the Scientific Revolution. This will close the circle by pointing both to the aptness and the insight of Réaumur's language metaphor with respect to his claims about making comparable thermometers.

RÉAUMUR'S DEVELOPMENT OF ALLEGEDLY
COMPARABLE THERMOMETERS

Réaumur's work came after a century of efforts to build comparable meteorological instruments, mostly baroscopes/barometers and thermoscopes/thermometers.[2] Comparable instruments would allow different observers to measure the spring of the air and its degree of heat at different places and different times, and the ensuing measurements to be juxtaposed so as to compare the state of the atmosphere in different locations and determine its evolutions. In the early eighteenth century, the Royal Society in London and the Académie Royale des Sciences in Paris had each committed a member (Maraldi in Paris, Derham in England) to gather observation journals from credible correspondents (thus involving networks of distant philosophers in the Republic of Letters), to collect and compare them.[3] This was a tedious business that involved information practices that would call for a detailed historical study of their own. Derham for instance labored to convert the local length units through which local philosophers read the variations of the liquids in their instruments, gradually evolving a comparative tabular format to present data collected in different countries.[4] In 1725 James Jurin of the Royal Society exceptionally published a text in Latin (to reach a wider international audience) proposing a way of performing measurements and organizing observation logbooks.[5] The comparative tables were then published in the journals of these institutions, mostly on a yearly basis and typically with a three year delay between the year in which the measurements were taken and that in which they were published. Simultaneously, both institutions engaged in an empirical research program dealing with the making of comparable instruments, with an increasing focus throughout the first half of the century on the issue of comparable thermometers; Amontons and la Hire in the 1700s proposed various instrumental designs. Their efforts were much reported and were publicized in Fontenelle's topical yearly synthesis of the important work performed at the Academy.[6] Hauksbee in England embarked on a similar program, culminating in the 1720s when the Royal Society sent several of his thermometers to its mostly northern and central European correspondents for a joint campaign of allegedly comparable meteorological measurements. Jurin's proposal was of course closely linked to that effort at thermometric coordination. To complete this rough overview, one must note that in the French case, the meteorological observations at the Academy were performed by the astronomers of the Observatoire. The struggle for comparable instruments and coordinated measurements over geographical distance must therefore be placed in the context of similar efforts by the astronomic community. Astronomers all over Europe engaged in the coordinated observation of celestial bodies (the much publicized and contested observation of the 1664 comet being a case in point).[7] French astronomers strongly committed themselves to a long-term geodetic program (drawing the map of France, determining the shape of the earth)[8] in which coordinated astronomical and geometrical measurements were held to provide more accurate maps and therefore to be able to displace traditional cartographic practices.

What Réaumur proposed in 1730 was a method to build thermometers that relied on two specific steps: first, a quantitative assaying of the wine spirit used in his thermometers, which significantly departed from traditional trials founded on sensory perceptions (such as the smell left over after the distillation of the liquid or its capacity to detonate gunpowder in a burning mixture); and second, a method to graduate the scale of the thermometers between two fixed points, that of boiling water in the upper scale, and that of a mixture of ice and water in the lower scale. These two methods allowed him to propose his "Rules for building thermometers of comparable degrees" in print for other philosophers to read and replicate, and it was indeed read by many as providing a reproducible process. Debating with Michelle over the respective merits of wine spirit thermometers (Réaumur's process) and quicksilver ones (in the Fahrenheit instrumental tradition) Christin remarked that Réaumur's rules indeed permitted the construction of thermometers that would always mark a given degree of heat in the same way, and (when properly put into practice) to have equally precise thermometers made everywhere.[9] However, Réaumur did not loosely rely on the skills of his contemporaries to replicate his thermometers, but actively engaged in making them. The Abbé Nollet, acting as his hired laboratory assistant in the 1730s, made a number of thermometers that Réaumur disseminated in a network of acquaintances. This included philosophers such as Christin and provincial learned academies, but he also provided thermometers to a group of travelers of particular relevance to our story: his fellow academicians who were painfully performing geodetic measurements on the famed Peru expedition. Others included bureaucrats and travelers involved in holding diplomatic outposts or explorations such as Taitbout, consul at Algiers; Granger, exploring Cyrenaica and later the Middle East; de la Nux in the Isle de Bourbon; and sailing officers and doctors (Cossigny in the Indian Ocean, Poligny all the way to the Cape Verde Islands and back, Artur in Guyana). Réaumur then collected the varied sets of observations made by this traveling network and published yearly comparative tables in the *Memoirs* of the A.R.S. throughout the 1730s. The justification of the whole enterprise was in itself twofold, because it involved both measuring the striking degrees of heat that could be reached in the torrid zone and assessing the various degrees of heats that could be experienced by French expeditions abroad in order to better prepare them. The former argument pointed to the tradition of gathering curious facts and the latter to that of assembling useful data. Characteristically, Réaumur described his thermometer as simultaneously "amusing and useful."[10]

Still, Réaumur's phrasing pointed toward the significant displacement of several sedimented layers in instrumental practices. As he himself remarked, making his instruments involved a lot of labor and skill in order to make them comparable. Performing comparable meteorological measurements with them implied disciplined daily observations and Réaumur was sensitive to the fact that it might appear improper to require such discipline from some gentlemen. Some of his observers went to great lengths to satisfy him: Granger used to tread the desert paths of Syria with the thermometer suspended between his legs in a pouch, protected from the sun by the shade of his own body.[11] Implicit in his criticism of the fact that one only understands a thermometer one has used for a long time is the idea that comparable thermometers can be used by any skilled observer. In that case the ancient link between the natural philosopher and the tools of his trade is severed. Where one used to speak of the instrument of Mr. so-and-so who had long been using it, one now dealt with an anonymous commodity. Where credible measurements had united the credit of the philosopher and the quality of the instrument into a singular familiarity deriving from long-term practice, almost any observer could perform credible measurements on the spot with a proper thermometer.

Whereas an instrument could often be exchanged in a gift-giving credibility-swapping economy in the Republic of Letters, it tended more in Réaumur's scenario to become a standard commodity to be exchanged for money between "ordinary" practitioners. The moral economy of Réaumur's thermometers is thus no longer that of Galileo's telescopes, pointedly given by the latter to high-ranking princes, or of the Abbé Bignon's barometer, which stood in his private cabinet and whose erratic behavior was a cause of great concern for the Paris Academy in the 1700s.[12]

Another displacement dealt with the representation of distant places. For Derham or Maraldi, who were committed to comparing meteorological observations performed in different places, measurements had to be decontextualized from their context of production, and recontextualized in comparative tables: this recontextualization was very much performed through the tabular graphic rationality of their display, particularly before a consensus had been achieved over the feasibility of comparable barometers or thermometers. Allegedly comparable thermometers (and barometers) also allowed this representation of the meteorological conditions far away to be inscribed on the instrument itself, on the gilded wooden support where its scale was also drawn. Michelle had proposed to draw a mark on his thermometer indicating the height of the quicksilver in the height of Pondicherry or the cold of Torcea in Lapland.[13] The instrument therefore came to condense the network of instruments, observations, and philosophers, woven together through the comparability of their tools. Himself one of Réaumur's allies, Christin approved of such practice. He judged that indeed "the thermometer was a kind of geographical map, where the various degrees mark the relationships between various known and unknown countries in the world. . . . Each day one is delighted to know how far one stands from the degree of heat in Sumatra or that in Pondicherry," and that one could not add too many such comparable geographic terms on one's thermometers.[14] What was at stake here in the map metaphor was indeed the ability of the cabinet philosopher knowledgeable enough in instruments to share the experience of his traveling peers in different lands. As a geographer of the time commented, geography (a large part of which involved map-making in his mind) "paints the earth to us in reduction, it brings closer to our eyes countries where we could not walk, it gets us to travel at no cost and in the absence of danger, it somehow naturalizes us in all known lands and makes us citizens of the whole world."[15] We will later return to this issue of shared experience in relation to language. Before that, it is important to assess critically the comparability of Réaumur's thermometers.

As numerous studies on the standardization of instruments have shown, precision travels poorly.[16] The amount of labor needed for de-contextualizing measurements performed far away into data available for endless textual reappropriations is enormous. The case of Réaumur's thermometers is no exception. The letters of Réaumur's correspondents were full of remarks about the difficulties they encountered. In Guyana, although he checked them before leaving France, Artur was faced with three Réaumur thermometers indicating respectively 18, 20, and 27 degrees. Cossigny mentioned that his own two thermometers were different by two degrees. But however potentially damaging to Réaumur's claims, none of these comments went further than private letters addressed to him. One reason for this is the fact that this group of observers were Réaumur's allies. Artur used Réaumur's language metaphor in a way that was both teasing and a token of familiarity between the two men: "I could not neglect to remark here that my two thermometers do not speak the same language. The difference is too considerable to be left unsaid." Another reason was the ambiguity of some of Réaumur's claims. Even if he framed a horizon of exact comparability for instruments made according to his rules, he did not believe that goal to lie within reach of actual practice. Taking a pragmatic stance on the comparability issue, he seemed quite content to argue that one can make thermometers that differ little enough to give us an idea of degrees of heat that will be sufficient to one's needs.[17]

THE LANGUAGE OF INSTRUMENTS OR MEASUREMENTS
WOVEN INTO LANGUAGE

Réaumur's use of the language metaphor to sustain his claims about the comparability of thermometers made according to his principles was two-pronged: (1) different instruments could speak the same language, and (2) any observer was in principle capable of understanding the language of any instrument, not only that of his own, familiar by long-standing practice. Therefore the metaphor explicitly pointed toward the ability of a privileged and prominent observer, given wide access to meteorological data taken with Réaumur's thermometers in many places, to build comparative tables whose lines and columns could easily be permutated, for each set of data taken in each place weighed about the same. I want now to dwell more closely on what Réaumur's language metaphor left implicit, which was revealed in travel accounts describing meteorological measurements made with allegedly comparable instruments, whether barometers or thermometers.

When we look at the list of travelers who were given or sold Réaumur's thermometers in the 1730s, the Peru expedition looms large, both because determining the shape of the Earth by astronomical/geodetic measurements was a key issue in natural philosophy at the time, and because they continuously used meteorological instruments.[18] The account of the expedition written by the Spanish officer who accompanied the French academicians, Antonio de Ulloa, is particularly interesting in this respect. He described their tedious climb to Quitó from the sea through the Cordillera in a logbook format that ran in the following vein: "The 17th at six in the morning the thermometer marked 1014 1/2 at Taxoguagua [he goes on to describe a few events]; the 18th at Cruz, at six in the morning, the thermometer marked 1010."[19] And so on, day after day, with decreasing thermometric figures as they trod onward up to Quitó. This daily decrease, reembedded in the travel logbook format, provided a striking narrative rendition of the ascent. That rendition rested on two pillars.

First was the buildup of a sympathetic feeling in the reader through the embedding of allegedly comparable measurements in the account of distant experience. Through the mediation of measurements, readers supposedly familiar with the indications of Réaumur's thermometers (even if only in their own surroundings and/or through their readings) could sympathize with the academicians' experience of cold in the mountains. In a sense the measurements in the account, because of an assumed intrinsic comparability, allowed the armchair reader an empathic access to that distant corporeal experience. To make the point clearer, the academicians themselves argued for a symmetrical thermometric measurement-mediated empathy. Bouguer thus noted that using Réaumur's thermometer "we could look upon ourselves in Quitó as in France in spring or at the end of autumn."[20] Second was the textually staged continuous sequence of numbers in Ulloa's account of the travel to Quitó, which built on that measurement-induced sympathy as a narrative resource to convey to the reader a sense of sharing the experience of climbing as a continuous process. This was no small feat, for there were remarkably few ways (if any indeed besides saying it explicitly) for travel accounts in the early modern period to construct an experience of climbing that could be shared with the reader.

To take the example of a philosophically oriented account, De Thou's rendering of de Candale's Pyrenean ascent, the literary tools that make for an impression of the heights sequentially reached were the ability to climb higher than any trace left by human presence, higher than the clouds, at heights where only eagles could live, and, still above that, heights where the air got to be rarefied and subtle, and where the intrepid climbers would meet various afflictions such as exhaustion and dizziness.[21] Those were the main characteristic tropes of the times, with an additional one usually encountered at the top, which involved various modes of spiritual, moral, or aesthetic epiphanies.[22] Such literary tropes may describe stages of the climb, but except by their juxtaposition of discrete moments, they were awkward in rendering the ascent as an ongoing activity. There seemed to have been some mounting dissatisfaction over this ineffectiveness of travel accounts in the philosophical community, particularly when they tried to extract stable data from them. At the end of the seventeenth century, trying to deduce the height of the Teneriffe peak from a learned gentleman's account, Boyle remarked somewhat disappointedly that the estimate of the distance they had to walk (21 miles) gave a poor idea of the heights reached, because of the "crookedness" of the paths one had to follow.[23] Things started to change in the 1720s and 1730s when mountaineering philosophers started to carry barometers and thermometers made and handled according to rules that allegedly allowed their users to extract stable data referring to their environments, such as the height of mountains or the degree of cold in a particular place at a given time.

Ulloa's account of the Peruvian academic expedition is particularly interesting because it went one step further than providing data that could be circulated. By providing the temperature day by day, so that the daily drops measured with Réaumur's thermometer beat the rhythm of their ascent toward Quitó, he tentatively gave his readers access to the feel of their ascent as an ongoing process. But the quality of that feeling, the way it could be translated into the reader's own frame of reference relied on comparable thermometers. There would be no other available way to convey that continuous experience in a meaningful manner, for indeed "ordinary" perceptions, that is perceptions unmediated by data taken with comparable instruments and thus able, at least in principle, to be exchanged across the great divide of the author and reader's experiences, were not trustworthy. Ulloa remarked that when ascending or descending in the mountains, the traveler would respectively overestimate the feeling of increasing cold or overestimate that of increasing heat (with respect to some reference temperature, taken of course with Réaumur's thermometer). We could then be confronted by the curious spectacle of two travelers at the same spot on the mountain but dressed for different climates. Therefore, Ulloa or other authors could not testify to the relevance of their own feelings during the ascent, and an important avenue for conveying experience and asserting empirical claims to distant readers, namely, the witnessing technology, was thus closed. In the latter, the author either testified to what he experienced through his senses or he took the slightly more devious way textually of making the reader a virtual witness of what the author had perceived through circumstantial descriptions. I will return later to the difference between data-mediated sympathy/empathy and the use of the witnessing technology to ascertain empirical new and curious philosophical facts. The interest of Ulloa's account lies in his very sophistication. Not only did he give his reader a feel for the continuousness of their ascent, but this feeling could be translated in the reader's sense of experience because of the use of comparable thermometric measurements, similar to the ones the reader could be expected to have encountered directly (if he himself owned Réaumur's thermometers) or indirectly (through exposure to the philosophical accounts that gave and used figures taken with such instruments).

Although there was a great deal of concern in philosophical circles in the second half of the eighteenth century over the importance of sympathy (one can think here of Diderot's reflections on acting and of the place it occupied in Hume's and Adam Smith's works), sympathy was not (and still is not) necessarily mediated by data taken from instruments. In the culture of *sensibilité* that was pervasive in France at the time, being exposed to the spectacle of a virtuous act would make the spectator happy, because emotions and virtue were deemed to be close. He would not only become aware that a virtuous act has been performed, but also he would somehow empathize with it. For instance, the culture of historical paintings of the time thus rested on the notion of the exemplar.[24] By vividly depicting an exemplary act of virtue or dignity, usually taken from a common stock of (mostly Greek and Roman) antique and mythological deeds, the painter could count on inducing a controlled moral and aesthetic emotion in the viewer of his painting. The *exemplum virtutis* then takes the place of the temperature measured by comparable thermometers: it made for a common ground within the semantics of language from which to mediate between the intentions of the painter and the experience of the viewer, because it was supposed to speak to both in the same way. Although it would take too much time to pursue these approaches to the sympathetic mediation of distinct experiences in different fields, they are worth mentioning both because they were contemporary, which suggests the interest of a broader cultural approach, and because they relied on moral exemplars, showing that measurements were not the only available channel to intersubjectivity. But in both cases a shared reference point in language was needed to mediate between distant experience, whether a consensus over the comparability of thermometers or shared *exempla virtutis*.

Returning to the issue of constructing experience-mediating narratives through comparative measurements, I would like to take another example, namely, the balloon ascents performed by the French physicists Charles in 1783 and Biot and Gay-Lussac twenty years later. The scientific times were different in the sense that the comparability of thermometric and barometric measurements was no longer problematic. However, their use for accurate determination of altitudes was a more contested issue. Nevertheless the problem confronting the balloonist-physicists was similar to the one encountered earlier by Ulloa, and perhaps even more acute: there was no way (the cloud layer being a limited exception) for them to build a heroic narrative of their ascent outside the use of meteorological measurements to provide a sense of height to which the reader could relate. Instruments were central from the start, for these famed ascents were highly publicized as oriented toward physics because of the use of instruments, and, in the case of Biot and Gay-Lussac, one could say the ostentatious use of a whole battery of quantitative instruments and experimental devices.[25] In Charles's ascent, barometric heights were relied on to build for the reader some sense of what is going on, in a way that was similar to Ulloa's textual strategy: "The barometer was then oscillating at 26 *pouces*. We had stopped climbing, meaning we had climbed to about 300 *toises*. Since that time we had always followed a horizontal path, between 26 *pouces* of quicksilver and 26 *pouces* 8 *lignes*."[26] The narrative voice switched back and forth here, from the barometric figures to "human" estimates ("about 300 *toises* above the ground") to create empathy: men and instruments were deeply entangled. This entanglement threatened to become even more complex when the narrative got to the point at which the balloon reached its apex: "When the thermometer stopped climbing, I measured very exactly 10 *lignes*. This observation was of the utmost rigidity, the quicksilver suffering no visible rise; I deduced from that level a height of 1529 *toises*, waiting until I could integrate that calculation and put more precision in it."[27] The balloon was indeed followed by several observers who cross-checked its path with astronomic measurements, and the barometric heights were converted after its landing into calculated heights (usually indicated at that time by the multiplication of significant figures and the absence of rounding-off procedures). These calculations were made by the engineer Meusnier, who used the formula of Deluc and various adjustments.

This also occurred in the later accounts by Biot and Gay-Lussac. It was self-evident to them that they should carry many meteorological instruments, thermometers, and hygrometers.[28] Moreover, Gay-Lussac was keen on finding a regular variation law for the behavior of the quantities measured by those instruments with the altitude reached by the balloon. One of the motives for the trip could be found in a previous attempt by Robertson who reported a decrease of the earth's magnetic field, which the Frenchmen wanted to put to the test. With respect to other instruments, Gay-Lussac was disappointed in finding a "singular law" for the hygrometer and satisfied himself with excluding anomalous thermometric measurements in order to obtain a simple rule stating that the temperature decreases one degree every 173 meters.[29] The thermometric measurements thus became a new channel to render the experience of the ascension to the reader. Biot and Gay-Lussac would certainly not rely on their own sense of heat or cold; they stressed the most striking discrepancies or correspondence between these experienced sensorial perceptions and measured temperatures. These were enough to disqualify our perceptions of heat, without need for any further comment. The description of the heights they reached typically involved the temperature at $-9.5°$ the barometer at 32.88 cm and the height at 6,922 meters. What can be made of the various heights provided by Gay-Lussac? As in the case of Charles's ascent, those were obtained through the coordinated efforts of the men in the balloon, several observers simultaneously taking measurements on the ground, and a posteriori calculations by an engineer of the Ecole des Ponts-et-Chaussées, combining the empirically determined quicksilver level and the new Laplace formula. But contrary to the case of Charles twenty years earlier, these reconstructed altitudes were unproblematically given in the narrative itself. The maximum altitude was even given in relation to sea level, not only as a way to provide a direct empathic bridge between the author's and reader's experiences, but also to allow the latter to construct an indirect one and compare this figure with the mountain heights and travels he knew of. Measurements were meant to make sense to a reader where the testimony of the senses of the observers failed, whether for lack of milestones or because of their intrinsic unreliability; however, those measurements given in the account were determined after the fact by the coordinated activity of numerous people. The use of measurements made with comparable instruments therefore enabled a specific literary possibility for the reader to empathize with the author's extraordinary experience, and it involved a significant departure from the traditional witness account: the credibility of the figure did not rest in the senses and instrumental skill of the narrator, but rather on the coordination of a network of practitioners that extended farther across time and actual and social space than the event it contributed to depict.

CONCLUSION: SENSES AND MEASUREMENTS, OR THE SYMPATHETIC READER

In their seminal analysis of seventeenth-century natural philosophy, Steven Shapin and Simon Schaffer aptly emphasized the part of testimonies in the legitimacy build-up of experimental philosophy. Experiments were public displays whose account relied on the testimony either of credible eyewitnesses who were explicitly named or on the narrative of the philosopher that, by the stress he put on describing extensively both what he saw or felt and the circumstances that surrounded the experimental trial, made the reader a virtual witness of what happened. In this particular textual strategy the testimony of the senses was required, among which vision was foremost.[30] Although the degree of heat or cold could be perceived by the senses, and the subtlety of the atmosphere in the upper mountains could somehow correspond to some sense of uneasiness in the body, these perceptions ranked far below vision as a foundation for testimony. That is part of the reason why the case of the comparability of thermometers and barometers and the resources it provided for new avenues of persuasion or sympathy is revealing: these struck directly at the sensorial underbelly of the witness rhetorical strategy. Reliable instruments could more easily be accepted as mediators between men and nature in cases where the senses provided notably untrustworthy information. Comparable instruments could then act as mediators between man and man through the entanglement of narrative and measurements. Thermometers and barometers were not exactly alone in that respect, either. Instruments supplementing the senses such as the telescope and the microscope had lain at the core of the Scientific Revolution. Comparability of instruments and measurements had been a contemporary issue in other fields of enquiry such as astronomy and its application to geodetic measurements and cartography, and the quest for comparable meteorological instruments owed much to these parallel empirical approaches: in France for instance, the astronomers of the Observatoire were responsible for performing, collecting, and publishing comparable measurements. Because the temperature of its caves seemed to be independent of seasonal variations, the Observatoire provided the privileged locus where thermometers could be compared and tested. It thus operated as a frame of reference for Réaumur's own trials. But unlike celestial objects, the degree of heat or the spring of the air could be felt, however dimly or imprecisely, which meant barometers and thermometers could be made to relate to forms of bodily experience. Because of that very fact, the comparability of instruments and measurement could lead to new developments in textual practices. In travel accounts involving both meteorological observations and issues of trust in accounting practices, comparable measurements performed with comparable and widely available instruments (at least among the lettered elites) could become a resource on which to build a particular sympathetic channel in which the author's experience made sense to the reader's.

At this stage Réaumur's language metaphor thus appears as a particularly insightful one, maybe more in what it left unsaid, than in what it purportedly addressed. The issue was not only about instruments speaking to one another, or philosophers being able to listen to or understand any instruments beyond their own, it was also about men relating to other men through instruments. Dealing with comparability, it dealt even more with intersubjectivity. And intersubjectivity in the realm of meteorological experience meant speaking in comparable numbers, taken with comparable instruments, as Delille (himself a champion of comparability with respect to meteorological instruments) emphasized: "But to fix the imagination on the degrees of these various colds and those on which I will report later, it is necessary to express them with numbers, which one can do according to the temperature degrees of Mr. Fahrenheit, or those of Mr. Réaumur, or eventually mine."[31]

The sociologist Anthony Giddens saw the roots of modernity in the mechanism of de-localization of action and experience, through symbolic means of exchange that could be objectifiable and the delegation of trust to objectivized expert systems.[32] I would like to argue that what we see here was a particular instance in the build-up of modernity in the sense of Giddens, with the network of observers handling trustworthy comparable instruments as the expert system, which indeed led to a particular symbolic circulation: thus measurements produced in a particular context could be extracted from it and brought to bear on similar measurements performed in distant places by juxtaposition/comparison in narratives and tables. Measurements in this context already acted as uncoded information that could be circulated in proper textual vehicles and submitted to various forms of rationalizations and calculations. But we also saw that in the process, instruments, measurements, and language got entangled so much as to provide resources in the construction of intersubjectivity, or more precisely for conveying experience to distant actors. Modernity, as the case of Réaumur's comparable thermometers shows in a particular context, also means renouncing the taking of language for granted and looking closely at this entanglement of language, meaning, and practice that is a pillar of the social order as we experience it.

CHAPTER SEVEN

Language in Computing

Jörg Pflüger

La langue est à cheval entre la nature et la culture.
—Roman Jakobson, "La langue est le moteur de l'imagination"

There is no doubt that computer science is quite different from the natural sciences. I would like to characterize it as a peculiar amalgam of applied mathematics, the art of design, and social technology.[1] Computer science deals with modeling, and its three constitutive disciplines correspond to the representation, the execution, and the effect of such modeling. From this characterization, it follows that in computer science, language phenomena vary much more than in the natural sciences and mathematics.

Since computer science operates solely with formal languages and everything it produces is a formal artifact, it seems possible to insist that its production only involves mathematics and has nothing to do with language.[2] Such a statement is similar to saying that everything that can be programmed, can be programmed on a Turing machine. Although both these statements are correct, they have little significance. Different computer languages, although equivalent with respect to computational power, have nevertheless different expressive powers—a notion that cannot be grasped in formal terms because it refers to human powers of comprehension. In much the same way, formal constructs like programming languages and programs are social constructions as well. Their shape, development, and use are linked to their contemporary ideas about how human beings comprehend the world and communicate with each other. Formal languages are thus more than formal languages. They are a medium, in that they capture images of the world and mediate between human beings and artifacts.

Talking about the language analogy in computer science only makes sense if we distinguish between computer science's specific artificial languages, on the one hand, and natural language as the genuinely human medium of understanding and communicating, on the other. Thus, we perceive the formal constructions as reductions of "living" mediation processes. In itself, a formal language is nothing but a special mathematical calculus, an algebra with an associated grammar that provides rules for the production or analysis of its expressions (to which can be assigned "meanings"). However, it is not very rewarding to study structural similarities between such algebraic "rewriting systems" and static descriptions of natural languages in linguistics.[3] In order to speak about a language analogy within the concepts of computer science, it is necessary to go beyond the formal properties of formal languages and to consider the many facets of their uses in relation to human use of natural language.[4]

For a sufficiently comprehensive definition of language I will go back to Wilhelm von Humboldt's theory of language. "Language, regarded in its real nature, is an enduring thing, and at every moment a transitory one. . . . In itself it is no product (Ergon), but an activity (Energeia). Its true definition can therefore only be a genetic one. For it is the ever-repeated mental labor of making the articulated sound capable of expressing thought." This definition does not distinguish between speech and language; it implies that the "life" of a language is in its use, that its life is inseparable from human activity. "We must look upon language, not as a dead product, but far more as a producing, must abstract more from what it does as a designator of objects and instrument of understanding, and revert more carefully, on the other hand, to its origin, closely entwined as it is with inner mental activity, and to its reciprocal influence on the latter."[5]

To carry out Humboldt's program requires analysis of language from different perspectives and on different levels. Studying grammar leads to a solely static view that does not take into account a language's living character. "The break-up into words and rules is only a dead makeshift of scientific analysis."[6] If one analyzes a formal language as a mathematical structure, then this is the case. It is characterized precisely by the fact that its expressions can be synthesized unambiguously from elementary parts. However, the designative function of language serves to express thoughts and to give them a lasting form that informs further intellectual activity. "Language is the formative organ of *thought*." Or, "[Language] precedes the ordering of thought in inner and outer discourse, thus determining the mode of connection of ideas, which again affects man in all directions."[7] In this sense, programming languages provide a medium to give a necessary precision to formal modeling, while their repercussions on the ordering of thought is mirrored in a certain programming style.

Although a natural language indeed exists independently of individuals, it nevertheless lives only in the act of speaking. Through its usage, its formative power affects not only the ordering of thoughts but also its own formation. It survives in a feedback loop—in today's jargon—and thereby transforms itself. Humboldt wrote:

> We need to connect the representation of the construction of the methods of language with general examination of the influence of language on the intellect and on man, and, as since it is the living man who is in fact the sole true bearer of language, which can only be momentarily embodied in its passage through the mental faculties, so its reciprocal influence on him acts also again on the whole of language.[8]

A language influences how the world is seen and modeled, and the experiences of a speech community contribute to its further shaping. In contrast to natural languages, however, which manifest (only) an immanent history, it is difficult to modify artificial languages, while they more easily can be created for specific needs. The history of programming languages and interfaces therefore mirrors the experience of the intellect in devising formal models in all kinds of variants and ever-new conceptions of formal languages. (See Figure 7.1.)

The history of computer science has seen a proliferation of languages. All possible systems of notation and operational interfaces are called languages, such as programming or dialog languages, hardware description languages, query languages, or data definition languages. Rather than inspecting this Tower of Babel, I will concentrate on programming, or, more generally, on modeling with the computer; other types of languages are either no more than simple systems of formal notation or display similar characteristics to programming languages. As far as modeling is concerned, the language analogy involves, on the one hand, the leading ideas about programming languages and conceptualizing with higher-level languages.[9] On the other hand, it concerns concepts of interaction with the computer, where the user who solves problems is for the most part a nonprogrammer. The question, then, is how concepts of language have had an influence on various user interfaces.

The language analogy appears in quite different contexts of modeling: in the process of formalizing real world problems; in communicating such

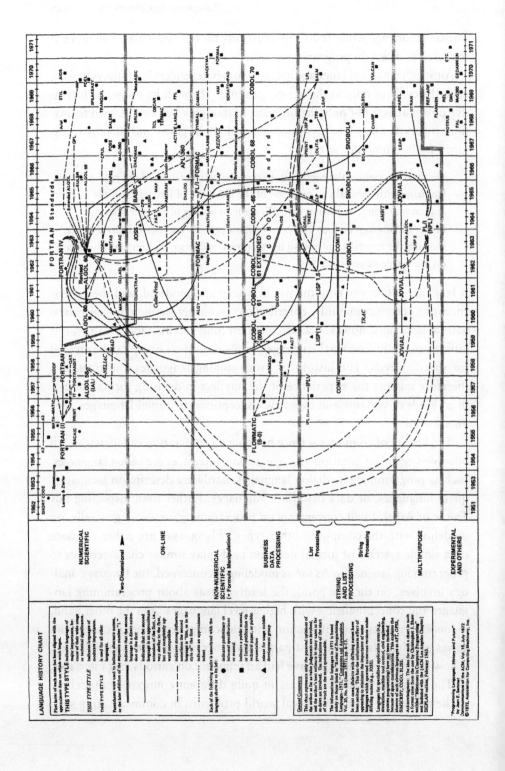

LANGUAGE HISTORY CHART

"Programming Languages: History and Future"
by Jean E. Sammet
Communications of the ACM, Vol 15, July 1972
© 1972, Association for Computing Machinery, Inc.

formalizations; and in various modes of interaction between human and machine, which acts as a special kind of "interpretant." The language analogy has played different parts in the short but eventful history of computer science. These roles depended, perhaps more strongly than in most other scientific fields, on currently dominant understandings of the functioning of (natural) language. The formal artifacts of computer science reflect different ways of dealing with language and different modalities of language in dealing with the world. Aspects of language incorporated into dominant conceptions of "correct" modeling have changed significantly over time. These changing views on language parallel those in contemporaneous linguistics and semiotics. In his short essay, "The Imagination of Sign," Roland Barthes points out on a very abstract level that one can distinguish three periods in which different aspects of the relation of signs were of main interest.[10] In the first, the relation between signifier and signified was taken rather naively as given; the second phase, which coincided with structuralism, took a "paradigmatic" view and concentrated on the systematic relations between signifiers; and the third and current—we might now call it "postmodern"—view of signs focuses on the "syntagmatic" process of producing them. More or less, these periods run parallel in time with the phases of the development of programming techniques described later. I think this semiotic shift characterizes rather well the epistemological background in which the history of concepts in computer science (and in other areas of knowledge as well) has taken place. Thus, the changing aspects of language in computing can be seen as a technological reflection of a more general epistemological shift.

In computer science, mechanical aspects of language are addressed as well as linguistic aspects of mechanization. Computer science deals with such different phenomena as formal semantics, automatic translation, so-called natural front ends, and computer-mediated communication. Its use of formal languages cannot be reduced to a technical terminology that solely expresses mathematical properties of formal systems; rather, their "artifactual"

FIGURE 7.1 *(opposite)* Jean Sammet, "Programming Languages: History and Future," *Communications of the ACM* (Association for Computing Machinery) 7 (1972): 601–10, 605. Timeline of programming languages up to 1971 courtesy of Jean Sammet and of the Association for Computing Machinery, New York. Jean Sammet's diagram shows the conceptual dependencies of the more important programming languages up to 1971. Today, such a history chart would be even more bewildering.

expressions entail new facts and realities. As computer technology deals not only theoretically with formal concepts but also reconstructs reality with its models, there exists a reciprocal effect between mechanization and language. Analogies thus work both ways. The artifacts of computer science affect, in turn, the realm of language, where the mind is at home, and they organize social reality in analogy with reductionistic models of language. One effect of computerization is that many activities are (re)structured in such a way that they seem to be organized by a "grammar of action" within a framework of formal languages.[11]

Drawing on Humboldt's theory of language, I attempt in the following to elucidate the development of various conceptions of formal languages in computer science from three different perspectives. These perspectives are aligned by their capacity to mediate between informal thought activities and operationally reified thoughts: that is, the way they permit "mentefacts" to be transformed into artifacts. These three perspectives raise the following three questions concerning modeling in computer science:

1. What is the purpose of the mediation of a formal language; or for whom is the formal representation of a "mentefact" meant?

2. What does the activity of modeling, supported by a programming language or an interface, look like?

3. How can we represent modeled reality by means of a particular formal language? What is characteristic of that model?

Of course, these three perspectives are closely interrelated, especially the second and third. They both consider how modeling takes place in computer science: first, with respect to the activity of the developer and/or the user, and second in view of the form in which their "inner mental activity" is expressed in a formal language. The way models are represented molds the modeling approach and vice versa: the experiences of the working process find expression in the means of labor and lead to the development of new programming languages. Both these aspects of modeling in computer science relate to the first question, since the relationship between human and machine, as mediated by formal languages, varies in the history of modeling.

Each of the three perspectives, can be organized in turn according to three views on the process of software development, in which the model of language appears differently. The threefold subdivision of the three dimensions reflects, more or less, a development in time or distinctive phases in the short history of computing. Introducing another trinity, one can observe in all dimensions a shift in stress from *syntactic* to *semantic* to *pragmatic* aspects of language.

1. Asking about the different aims of the mediating function of a formal language means looking for the primary addressee of a formal text. First of all, of course, it is the computer, which "interprets" a program and executes it. Second, in the course of the software development process, it is the developers themselves, whose ideas take shape in programming language or in an interactive dialog, and whose growing understanding is built on it. Third, with an advanced division of labor, the addressees are the colleagues with whom developers communicate about formalizable problems and their programmed solutions. The three addressees thus indicate the three different purposes of programming languages: they serve as a means to *instruct* a computer to do something; they help to *comprehend* (grasp, formulate, and solve) computable problems; and they provide a medium to *communicate* about algorithms and formal properties.

2. The history of software development and the history of software use can each be divided into three phases characterized by different guiding principles. In the first phase of both fields the dominant assumptions referred rather naively to language. This fixation on language was then overcome with a contrary concept of constructive activity, finally followed by a conception of communicative and performative action, which I would like to call "performative interaction," because it stages discourse combined with action in cooperative efforts of man and machine.[12] This change of paradigms, from *language* to *action* to *performative interaction* manifests itself both in the history of programming proper and in the interactive use of computers, although the phases of their development do not correspond in time.[13]

The successively dominant conceptions of how to produce software can be characterized by the metaphors *writing, building, growing*.[14] At the beginning, the task was simply to write a program; then it was to build it systematically, similar to the industrial mode of production; finally, the autonomous logic of the process of software development gave rise to the organic idea of growing software, thus taking into account the growing understanding and communication of developers. Concepts of interactivity underwent an analogous development, which can be expressed by the metaphors *conversation, manipulation, delegation*. Here, in the early days a linguistic idea also dominated, as interaction with the computer was conceived as a sort of conversation in a "man-computer symbiosis." With the arrival of PCs and computer workstations, the doctrine of "direct manipulation" became generally accepted, with the idea of handling knowledge by manipulating objects in a graphical interface. Today, in the time of the Web, the predominant idea is that in order to produce and supply information one needs to interact, communicate, and cooperate equally with software agents to whom one delegates tasks and with other human beings.

3. How a model can be presented in a particular language depends on the worldview embedded in the conceptualization of the language. Different features appear in history as the governing idea of a programming language or of interactivity that can be labeled: from *function*, to *representation*, to *agency*. Most of the early programming languages, and also the "conversational interfaces," were oriented toward the algorithmic proceeding and thus toward procedural or functional features. Later came object-oriented concepts, which have been trying to model observed or constructed entities of reality with the formal equivalents of "real notions" of the problem sphere. With agent-based systems the concepts of interaction and cooperation are incorporated in the program-world. They model distributed open systems through agents that can act autonomously and interact with one another and their environment. It is possible to describe the shift of the focus of representation also in linguistic metaphors: from *verb* to *noun* to *discourse*.

An illustration of these threefold perspectives might be helpful.

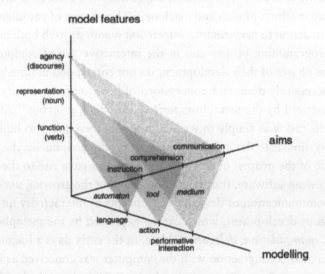

FIGURE 7.2

Looking at the cube spanned by the three outlined dimensions—disregarding that they are not actually orthogonal—one can lastly divide it along the main diagonal into three layers that correspond to the consecutive paradigms of the computer, guiding the evolution of computer technology as well as the public opinion of its users: the computer as an *automaton*, a *tool*, and a *medium*.[15]

INSTRUCTION, COMPREHENSION, COMMUNICATION, OR: THE MACHINE, THE HUMAN, AND THE OTHER

Probably the first programmatic claim for all three mediating functions of a programming language was made public in the proposal for the "universal language" ALGOL in 1959, where "universal" referred to the universe of numerical problems. ALGOL was to serve as a language that people could use conveniently to formulate algorithms in a form familiar to them, to publish and communicate about them, and these algorithms could be directly tested and executed:

1. "The new language should be as close as possible to standard mathematical notation and be readable with little further explanation."
2. "It should be possible to use it for the description of computing processes in publications."
3. "The new language should be mechanically translatable into machine program."[16]

These three different purposes of a programming language have not always been considered equally important. At the beginning, when programs were written individually for specific machines, and when the (numerical) procedures to be coded were mostly well-known, aspects of understanding and communication played no significant role. The primary task was to give the computer instructions, which were initially called "orders." During this period, programming was not thought of in terms of language in any emphatic sense. The usual term for a programming language was at first a "coding scheme," then with the arrival of higher-level programming languages an "automatic coding scheme." Of course, the expression "language" was common from the very beginning as well. However, it meant nothing different from the term *language* in mathematics. For example, Alan Turing used it in this way in a lecture to the London Mathematical Society in 1947:

> I expect that digital computing machines will eventually stimulate a considerable interest in symbolic logic and mathematical philosophy. The language in which one communicates with these machines, i.e. the language of instruction tables, forms a sort of symbolic logic. . . . Actually one could communicate with these machines in any language provided it was an exact language, i.e. in principle one should be able to communicate in any symbolic logic, provided that the machine were given instruction tables which would enable it to interpret that logical system.[17]

During these first years, programming did not mean much more than simply coding, with coding schemes figuring merely as a system of notation. John von Neumann was never convinced of the necessity for such a thing as a programming language. Algorithms were given in advance—mostly numerical procedures formulated in the "language" of numerical mathematics—and programmers were faced (only) with the problem of transferring the algorithms into a sequence of instructions to be executed by the computer. Formal semantics was no real problem, because the syntax of coding closely followed the anticipated operations of the machine.[18]

In the 1950s, computer scientists increasingly aimed at making programming more independent from concrete machines and developed artificial languages for more abstract machine models which were sometimes called "pseudo-codes." As the problems to be programmed grew much more complex, the focus of attention shifted from coding to the process of formulating and solving these problems. So-called problem-oriented programming languages appeared; the first one was FORTRAN in 1957.[19] Probably, the phrase "programming language" came into use with problem-oriented programming languages, because they conveyed more the feeling of dealing with a real language. The intention behind them was to support the formulation of algorithms down to the details on a more conceptual level. The computer was to take over the tedious and error-prone task of coding, giving human beings the time to focus on the problem instead of on the idiosyncrasies of the machine itself. The developers of FORTRAN wished to enable the IBM 704 "to code problems for itself and produce as good programs as coders (but without errors)."[20] Around that time, the perception of the computer also began to change: it was no longer considered only as an elaborate numerical calculator, but it was recognized that "the essential nature of the computer is that it is a symbol-manipulating device."[21] New types of programming languages emerged, like list-processing and symbol-manipulating languages, meant to open up new fields in a natural way. Victor Yngve introduced COMIT (1957–61) as a

> language that would foster fluency and freedom of expression. . . . COMIT is a user-oriented general-purpose symbol-manipulation programming language. It is user-oriented in that it is a high-level language that it is easy to learn and to use. . . . In many cases this meant the carrying over into COMIT of patterns familiar from natural language. COMIT is decidedly not, however, one of the programming languages that allows one to "program in English." But it is a programming language that takes advantage of intuitive feelings of naturalness that stem from the user's fluency in a natural language.[22]

Actually, it was claimed propagandistically that COBOL (which was developed between 1959 and 1961 and is the most important programming language ever in the commercial realm), allowed programming based on natural language, since one could use English words instead of mathematical symbols for the operations.[23] This evoked an absurd aspect of communication: this naturalness (of language) was supposed to give insecure managers the illusion that they could in fact control the work of their new wizards.

Problem-oriented languages require a compiler that automatically translates higher language features (source code) into machine code. This transformation represents much more than a mere transliteration because the structure of the two languages can be quite different. The more machine-oriented intermediate levels, which became a little more user-friendly in that it was possible to use symbolic names and macroinstructions, were then called (symbolic) "assembly languages." The construction of compilers required the clarification of concepts of formal grammars. The emerging theory of formal languages incorporated concepts of contemporary linguistic theories from Noam Chomsky and others, and perhaps the word *language* as well. With the more complicated translation of languages on different levels of abstraction, the problem of formal semantics arose. According to the focus on (abstract) machines, this formal semantics was conceptualized as operational semantics.

With the introduction of problem-oriented programming languages and ever more complex tasks, conceptual difficulties of program design moved into the foreground. The challenge was now to understand the problem and to find a programmable solution. Programming languages shape the way programmers acquire knowledge about formal systems and for that reason should be adapted to the conceptual level. The focus of attention shifted therefore increasingly toward the mediatory function of a programming language as the "formative organ of thought." Seen from this angle the history of programming shows a wide range of quite different types of programming languages, for example, imperative, procedural, functional, declarative, logical, rules-based, object-oriented, and agent-oriented programming languages, which I do not have the space to develop here. Each entails different styles of programming and suggests different modes for conceptualization. Many of these languages emerged in parallel, each designed for specific areas or uses. However, their history reflects increasing experience with programming, as weaknesses of earlier conceptions led to new ideas about design and new programming languages that were better suited to express revised procedures. The inner dynamic of a living language—that "in addition to its already formed elements, language also consists, before all else, of methods for carrying forward the work of the mind, to which it prescribes the path and the form"—can be seen in the artificial world of programming in a succession of design concepts and in the form of a Tower of Babel of fashionable, remote, or abandoned programming languages.[24] (See Figure 7.3.)

In so-called batch-processing, the computer is not integrated in the formation of understanding during the process of solving a problem.[25] The computer does not help in the conceptual treatment of problems, because it is used only afterward to process data according to predetermined procedures, which means that the programmer has thought the problem through in advance, so it is in principle already solved. However, there are many problems that cannot be thought through in advance and that are more suited for a trial-and-error process; there are even problems that cannot be formulated without computer aid at all. In those cases an interactive procedure, where the computer enters the stage earlier, seems more appropriate, leading to a "man-computer symbiosis," in the words of J. C. R. Licklider.

The idea of automating intellectual activities dominated the mainstream of batch-processing with the aim of freeing human beings from "lower" brainwork, thus giving them more time to concentrate on their "true" tasks. Commander Grace Hopper expressed conventional wisdom in an article from 1952 with the remarkable title, "The Education of a Computer": "It is the current aim to replace, as far as possible, the human brain by an electronic digital computer." Once routine mental activities are fixed in so-called subroutines and are accessible in routine-libraries, the user can turn again to more creative activities: "The programmer may return to being a mathematician."[26] Here the computer is perceived as an automaton that was to

FIGURE 7.3   Cover of *Communications of the ACM* (Association for Computing Machinery) 1 (1961). Courtesy of the Association for Computing Machinery, New York. The cover of the *Communications of the ACM* showed a Tower of Babel of programming languages in 1961. Although the early period was already a rather fertile one, it would be most tedious to redraw it today. Most of the languages have never come into real use or are forgotten.

replace human beings by working independently from them. The concept of interactivity introduced the alternative idea of a partnership between the human and the machine, a cooperation in which the human being is not replaced by the machine, but in which the machine supports the human being. As both contribute their respective capabilities to a dialog, a synergistic effect is to be expected such that "the cooperative interaction would greatly improve the thinking process." J. C. R. Licklider was the leading visionary of such a "symbiosis," funding research with ARPA money. He wrote in his famous article "Man-Computer Symbiosis": "The hope is that, in not too many years, human brains and computing machines will be coupled together very tightly, and that the resulting partnership will think as no human brain has ever thought and process data in a way not approached by the information-handling machines we know today."[27]

Although the dialog consisted in reality mostly of the exchange of commands and data, I believe nevertheless that interactivity established a new kind of reflexive comprehension, whose significance can hardly be overestimated. This new form of on-line thinking can best be characterized by a tendency to diminish the separation of the "inner" from the "outer discourse." Like in a conversation the development of an idea is chopped into pieces and guided by short-term reactions of an opponent and by the exchange of common externalized representations of thought fragments. "Inner mental activity" is distributed in the pauses between the short response cycles of the computer, and the "mode of connection of ideas" is governed by the turn-taking interaction with the machine. Thus, in interactive programming the first two moments of mediation—both of *direction* and of *self-communication*—are closely intertwined in a feedback loop. Using the answers of the instructed machine, the user instructs him- or herself how to think and to proceed further. "Always in full control, the user directs the mind-machine processes by commanding the computer to function, step by step, in accord with his deepening understanding of the problem."[28] The computer is thus thought of as being capable of assisting the human partner in his or her thinking, "amplifying his thinking" or "magnifying his mental sources." In this thinking by conversing with the machine its responses are as important as its responsiveness, and the "transmission of ideas" should be possible in a language designed for the convenient use of human beings. There was a lot of talk about a "conversation with a computer."

As early as 1956 Douglas T. Ross proposed the concept of "Gestalt programming," a form of *conversational programming* in which man and computer work together as a small problem-solving unit, man exchanging ideas with a mechanical interlocutor in a Gestalt language. "The purpose of a Gestalt system is to facilitate the transmission of general ideas as in a conversation, between a human and a computer, so that the maximum use of their respective capabilities can be made." The efficiency of this partnership depends on an appropriate language for the human: "If the human and computer are to work together to solve a problem, there must be some means provided for the transmission of ideas or results between the two, since the contributions of each will depend upon the actions of the other. . . . A major problem, then, in using humans and computers together is to choose an appropriate language for the interchange of ideas." In order to be suited for the human partner, it is necessary "to have the language operate entirely at the idea or concept level."[29]

Ross went so far as to assert that good cooperation requires a language "designed not merely for the purpose of communication, but for the convenience of fluent conversation."[30] His concept of language is remarkable in its simplicity; but we should remember that at this time many people believed in the possibility of automatic translation. "A language consists of two parts; a vocabulary and a set of syntactical rules. An idea is then transmitted by transmitting the expression of the idea; i.e., a sequence of words from the vocabulary. The final stage in the transmission is recognition by the receiver." Elementary words are attached to basic ideas and these words, following syntactic rules, are used to form more complex ideas, the Gestalten. "The topic for conversation is broken down into the finest logical divisions necessary to cover the entire scope unambiguously and with a minimum of rules for combination. . . . All of these various types of basic units will be called items, and the complete set of items forms the vocabulary of the Gestalt language. Thus a Gestalt is expressed in this language by combining items according to syntactical rules." This "dead makeshift" is indeed a very restricted code. The concrete form of a Gestalt system reminds one of a switchboard, where the "conversation" consists of pushing buttons and blinking lights. Ross himself used this analogy for his idea of a fluid conversation: "A closer approach to the ease, speed, and flexibility of a spoken language can be achieved by letting each word or phrase which is to be used be represented by a single unique switch or push button. A statement is then 'spoken' by pushing appropriate buttons."[31]

Although the concrete examples that were realizable at that time appear to be ridiculous, the dream of conversation with a computer remained alive and grew much stronger in the early 1960s with the interactive capabilities of time-sharing systems and dialog computers. There, the computer does not only compute a programmed solution, that is, an already solved problem, but also enters the process of problem solving in all phases from stating, formulating, and coding the problem to its solution. The dialog leads to novel and deeper understanding, interaction thus interpreted as an amplification of thinking.[32] "Through its contribution to formulative thinking which will be, I think, as significant as its solution of formulated problems, the computer will help us to understand the structure of ideas, the nature of intellectual processes." Licklider considered dialog computers a formidable instrument to form thought in a new and superior way: "Seriously, the man-machine symbiosis gives us a much better way than we ever had before for saying what we are trying to say and then finding out whether it is indeed correct. . . . I think the computer offers a real match to the problem of getting knowledge into human skulls."[33] Whereas Licklider stylized the dialog with a computer into a way of thinking suited to a modern world, its disciplinary moment is more clearly to be seen in Ross's statement: "Once this language has been designed the human is allowed to discuss the problem only in that language so that, in effect, a part of the programming of the problem has been accomplished by programming the human."[34] The dialog entails instructions for correct thinking and fosters—in a parody of Grace Hopper's phrase—"an education of the human."

The third aim of programming languages, that is, to come to a shared understanding about problems and algorithms, was originally, as in the proposal of ALGOL, thought of as providing a standard for the publication of algorithms. Subsequently, the increasing division of labor in programming produced many communication problems. Moments of understanding and communication seemed to be much more closely intertwined and the development of software had to take this into account.

> Source code not only communicates with the computer that executes the program, through the intermediary of the compiler that produces machine-language object code, but also with other programmers. . . . In most programming languages, far more space is spent in telling people what the program does than in telling the computer how to do it. The design of programming languages has always proceeded under the dual requirements

of complete specification for machine execution and informative description for human readers. . . . "Expressivity" became a property of programming languages, not because it facilitated computation, but because it facilitated the collaborative creation and maintenance of increasingly complex software systems.[35]

There have been various approaches to solve this problem of dual requirements, the most obvious one being to enrich programs with commentaries and documentations. However, none of these ever worked satisfyingly and much effort was spent on how to encourage or force programmers to take into account the readability of their programs.[36]

A radical, but rather late approach to tackle the aspects of comprehension and communication simultaneously was Donald Knuth's concept of "literate programming." "Considering programs to be *works of literature*," he introduced a programming language, called *WEB*, that allowed the programmer to write his or her program in a "stream of consciousness" and that was at the same time intended to be readable by his or her colleagues and not directly by the machine.[37] The machine should no longer force its syntactic structure on the programmer; rather, the programmer should be able to follow his or her own train of thoughts. The final program text is then automatically rearranged and "cleaned" so that the machine can "understand" it, too. Since a human being follows a train of thought more easily than a machine text, these programs are more suited to be read by other colleagues. An extensive commentary integrated in the program text should further enhance its readability. This conception of fluent writing was not widely accepted, presumably because it came too late. In the meantime, quite different ideas of a more systematic construction of software had prevailed.

Software development increasingly emphasized the constructive moments of design and therefore increasingly relied on higher-order conceptions of language, which would support modeling and not serve primarily for coding. The issue of communication came to the fore, no longer separable from the issue of comprehension, and not only on the level of readability, but with respect to the complex task of designing a program system in a distributed process. In the period of structured programming, in the 1970s, Douglas Ross himself developed the Structured Analysis and Design Technique, which modeled the world with little boxes and arrows. Although this bore little resemblance to a "fluent conversation," Ross nevertheless again called his method a "language for communicating ideas." However, instead of communication taking place between the user and the machine, the conversation was now between colleagues.[38]

A higher-order modeling language is to be used in earlier phases of the software development process before an implementation in a programming language takes place. It supports the agreement between the developer and the client about the form of a program system that is to solve an informally agreed upon task, as well as supporting the formation of its shape in the heads and the working documents of the developers, and it is used for communication in a development process constrained by division of labor. Today, in many realms, especially those where one has to deal with incomplete or imprecise knowledge, one goes a step further and integrates natural-language specifications during early phases of system design. It is an attempt to create a continuous transition from verbal communication about the task to be programmed, over the various mediation steps with their increasing formalization, to a formal specification, and finally to an executable program. In this stepwise process of mediation the primary addressee changes from the customer or colleague to the machine. In that way a cognitive process unfolds specific to computer science: from informal reality to its formal reconstruction.

All three mediatory aspects of formal languages appear with the spread of computer networks, the ubiquity of computer-mediated communication, and the rise of on-line communities in connection with interpersonal computer-mediated communication. The early time-sharing systems were thought of as libraries by their users, that is, public places where one could meet other colleagues, access information and tools, and get advice. These time-sharing systems had already led to user communities in which interaction with the computer and social communication and cooperation had been mutually stimulating. Early on, Licklider envisioned such a mixture of the two modes of communications (interpersonal communication and interactive thinking with the machine) in connection with computer networks:

> Most people's interactions with other people and with their intellectual surroundings will be mediated by networks. . . . Person-to-person communication will develop in such a way that it is routinely supported by access to data bases and models. Indeed, thinking will develop in somewhat the same way, being supported much more intimately than it now is by data bases and models. Interpersonal communication, which is essentially a thinking together, will derive much benefit from its mediation by an active network that will support it with processing and with memory.[39]

Today, it is possible to state more precisely, at least technically, how these two modes of communication are linked in the computer and how machine instructions are mixed with informative content: "And with the spread of the Web, which mixes human-readable text (HTML files) with programmed functionality in the form of CGI scripts, in-line JavaScript, server-side Java, or ASP, the boundaries of human-computer communication and human-human communication are increasingly blurred."[40]

Furthermore, communication in these electronic nets leads to cultures relying on shared interest in a programming language or an operating system like Linux, furthering worldwide cooperation, as in the open-source movement. Thus, a formal language constitutes not only a "speech community" of man and machine but also can function as a connecting link of a culture, based on mutually shared interests. Larry Wall, the developer of the programming language Perl, characterizes it in this sense as a "glue language" and states with respect to formal language the philological truism: "A language without a culture is dead."[41]

LANGUAGE, ACTION, AND PERFORMATIVE INTERACTION

I will now examine alternately the parallel lines of development of programming proper, on the one hand, and of interaction with the computer, on the other, to figure out what "correct" modeling meant in different times. These two histories can both be subdivided into three phases each, where the consecutive leading conceptions can be characterized by the succession of metaphors: (1) *writing, building, growing* for software development; and (2) *conversation, manipulation, delegation* for the interaction in the interface. Although the phases do not correspond in time within the two realms, they do have something in common in each phase. In both realms, the first phase was dominated by a rather simple concept of language that was superseded in the second phase by concepts of action, and the third, the current phase, shows a revival of language conceptions that combine discourse with action in a "performative interaction" reflecting the performance of the models homologous with that of their modelers.

During the early history of programming, the main approach to programming can be characterized as *writing* a program. The task was to formulate a given procedure for the machine. The difficulty of coding was regarded merely as a problem of notation; therefore a "moderately mathematically trained person should be able to do this in a routine manner."[42] When an algorithm had been sufficiently detailed, the real difficulty in translating it into machine code existed in the programmer's need to anticipate mentally the dynamic behavior of the machine and to structure it in a static scheme. Moreover, it was necessary to know well the individual machine for which the program was to be written. Since the capacity of the machines was rather modest, making efficiency the highest demand, the skill of programmers consisted in getting the most out of "their" machines. Thus, program texts were entirely written for perusal by the machine reader, and the quality of the writing consisted in the adaptation of the "text" to its only "reader." Good writing was based on "empathy" with the machine, which constituted an almost personal relationship, as Hopper's title "The Education of a Computer" indicates. Personal authorship was supplemented by large libraries of executable texts called subroutines.

This idea of "writing" did not change significantly with the arrival of the first problem-oriented—also called user-oriented—programming languages. Texts became less dependent on particular machines. The programmers wrote, so to speak, for an abstract model of a class of machines—a rather anonymous "readership." However, the organization of a program was still oriented toward its execution in steps or jumps, and good writing still consisted of "empathy" with the mechanical reader; the imperative or procedural higher-level languages (of the first generation) and their programmers had inherited the burdens of the von Neumann computer architecture.[43] The grammars of higher-level languages need to have a special structure so that programs can be parsed and compiled in real-time. Since the compiler was expected to produce a highly efficient machine code, the design of the supposed user-oriented languages oriented itself to the capacities of the mechanical translator and executor.[44] In order to meet both needs—of human and machine—computer science took a linguistic turn. To construct better compilers for more convenient use of languages, formal languages and grammars became central areas of research.

Under the paradigm of "writing," programmers did not conceive the constructs of higher programming languages as a means of structuring; rather, they perceived them as abbreviations of instruction sequences that facilitated the readability of a program text; and they used *procedures* as a means to save memory and writing labor, not considering them as conceptual units.[45] Computer scientists conceived of difficulties of comprehension as coming from the text and not from the organization of the design. Programming languages served to "say" something efficiently, and not to govern the "ordering of thought." Neither theoretically nor practically did computer scientists view programming languages as systems suitable primarily for furthering the organization of the design and representing it in an orderly fashion. One could say, perhaps, that in computer science Ferdinand de Saussure's distinction between *langue* and *parole* was not yet established; questions of performance determined the essential features of programming languages. Accordingly, semantics was conceptualized only as operational semantics, that is, the question of how something written will be executed on a model interpretant.

As discussed earlier, interactive computing in a partnership between man and machine formed an alternative current alongside the mainstream movement of batch-processing. A linguistic perspective likewise informed the conception of interactivity and the interfaces belonging to it. Batch-processing requires a complete and well-thought-out program text. "This vestige of the 'complete program' requirement forces the programmer to think ahead to the total structure of his program at a time when he would rather concentrate on translating a few simple ideas into program statements."[46] Inasmuch as interactive programming does not require this thinking ahead, the obvious thing would be to see in it a form of experimental or creative writing. In fact, however, the metaphor of a *conversation* prevailed; "real-time interaction" was synonymous with "conversational mode."

The computer was no longer to be an obstinate machine subject to educational efforts or to be an anonymous reader, but should hide behind the mask of an interlocutor. One of the first dialog languages was JOSS, implemented around 1962 on the abandoned computer JOHNNIAC (named after John von Neumann) and also used regularly by nonprogrammers. Its developer, Cliff Shaw, explained its design principle, meant to give the user a "feeling of linguistically directing an agent": "To carry out the philosophy of presenting JOSS to the user as a computing aide and the *only* active agent with which he communicates, it was necessary to 'hide' the Johnniac from the user and to present instead the image of a person interpreting instructions and remaining in control of the situation no matter how senseless those instructions may be."[47] However primitive such "talk" might have been, the machine's prompt responses gave the impression of a dialog.[48] "JOSS lacks the problem capacity to carry on a sophisticated conversation. For its comments, it simply selects from a stock of 40 'canned' messages. But the timeliness and appropriateness of its remarks give a feeling of interacting with a person."[49] If its users are to be believed, the illusion was successful: "The JOSS computer wasn't just a tool one would work with, it was more like a friend who helps you with your work. He's also polite."[50]

At the end of the 1950s and in the 1960s it was commonplace to talk about conversation with the computer. In early articles like "Conversation with a Computer" and "Computer Conversation Compared with Human Conversation" this was taken literally, showing an exceedingly naive belief in the feasibility of automatization of language.[51] Later, this idea was expressed somewhat ironically, like in the 1967 articles "Conversation with a Computer" and "The Professor and the Computer: 1985."[52] The original euphoria and fascination with a literally understood conversation vanished with the daily experience of working with a time-sharing system. However, the idea of conversation between a human and a computer took hold as a metaphor for the interactive mode of operation of amplifying thinking in a "feedback discourse circuit."[53] Expressions like "conversational programming," "conversational computers," and "conversational languages" were in general use. And the computer accessible in conversational mode was conceived as a helpful other, addressed by varying degrees of servability as "partner," "assistant," "intellectual servant," or "handmaiden."[54]

At the end of the 1960s, however, the analogy with language lost its prominent place in favor of a spirit of construction, structure, and action. The abandonment of the "linguistic turn" showed in many areas of computer science. One of the pioneers of computer science, Maurice Wilkes, noticed in his Turing Lecture of 1967 that "people have now begun to realize that not all problems are linguistic in character, and that it is high time we paid more attention to the way in which data are stored in the computer, that is, to data structures."[55] By the middle of the 1960s, at the latest, it became abundantly clear that the production of software could not be achieved simply by dividing up the writing tasks and that the planning of larger projects could only succeed with a certain standardization and organization of the programming labor. The famous NATO conference of 1968 made the "software crisis" public and established as a remedy the discipline of software engineering.

The idea of a "software factory" was a response to the software crisis. Replacing the often opaque work of individual artist-programmers, the work was now to be rigidly organized "as if one produces software in the same way that one manufactures spacecraft or boots."[56] With ever more complex objectives, the design itself—therefore the problem of understanding the problem—turned out itself to be the problem. "In short, it is the thinking errors, more than the coding errors, which limit programming productivity."[57] The task was now to break down the problem with a top-down strategy of *divide et impera*, so that the solution could be constructed from relatively autonomous modules and therefore based on the principle of the division of labor. The design was to spread out the logical structure of the problem; from its abstract formulation, the solution was to develop by "stepwise refinement" into a system of partial models in a treelike hierarchy of abstraction. The dynamic of software development was then to follow the anticipated structure of the solution; the order of the product should organize its production.

In the 1970s, this concept of *structured programming* was promulgated as the doctrine of good software production, and with it the metaphor of *building* a program. The term *building* connotes the composition of an object out of simple components as well as a rational design—thus describing the activity of the mason as well as the architect. Accordingly, in this double sense structured programming meant the syntactical composition of programs out of a few standardized elements, as well as the conception of a top-down design. Under the building paradigm, a programming language was to provide elementary features functioning as simple building blocks. With these it should be possible to construct an orderly program whose structure reflected as transparently as possible the relation between the static control structure and the dynamic process of execution. The orderly structure should make the program more easily readable and less prone to errors. The second meaning of building had more important consequences. This aspect was not primarily concerned with transparency of code, for finding errors or for further modification. Rather, a programming language was above all to facilitate oversight during the construction of a programming system. Under the paradigm of building programs the relevance of programming language as a coding scheme diminished compared to its role in supporting the specification and design of an executable system.

Theoretically, under this doctrine, the software development process was divided into two distinctive phases: analyzing the problem and synthesizing its solution. Before beginning to program, it was necessary to conceive of the problem, or the solution, as a whole and to formulate it "paradigmatically" in a formal specification. Therefore, in addition to programming languages proper, other types of languages (like specification languages) and other means for expressing more abstract concepts (like visual languages) became more and more important. The notion of construction, no longer oriented primarily toward the course of execution, suggested a different notion of semantics, too. So-called denotational semantics represents a building kit in which computable functions can be constructed from elementary semantical building blocks. Under the structural view, the theoretical focus of attention had shifted to the programming language as *langue*, in which everything depends on its place in a static and closed sign system.

The shift from the conception of language to one of constructive activity and tools reached the sphere of interactivity only a decade later, in the late 1970s. It did not run parallel in time to structured programming, but to subsequent object-oriented design techniques. Since interactivity proceeds in a flexible and experimental manner, a top-down approach makes little sense here. A type of user who employs tools first appeared with the incremental bottom-up handicraft-approach of object-orientation. This user came along with the idea that knowledge can be *manipulated*, by symbolic handling of visual representations of tools and objects of work in a "desktop" interface. The interactive activity concept that found its most pithy expression in the concept of *direct manipulation* was meant for a small "reactive engine" and for workstations, whose users were predominantly nonprogrammers.[58] These users did not intend to communicate with an opponent or to operate a machine, but to make good use of software tools in their own work. Interactivity neither continued to play a role within the innovative process of thinking nor served to augment it in a partnership. Rather, the sense of a manipulation-interface is "to provide computer support for the creative spirit in everyone," now emphasizing creativity as a playful or useful activity.[59] Where in the beginning was once the word, now it was action.

The idea of manipulating knowledge with tools required easily manipulable input devices and was to prevail only with the arrival of graphical interfaces.[60] Furthermore it required superseding the fixation on language so present in the idea of conversation. "Systems having direct-manipulation user interfaces encourage users to think of them as tools rather than as assistants, agents, or coworkers. Natural-language interfaces which are inherently indirect, encourage the reverse."[61] Characteristic of the concept of direct manipulation is that the symbolic activity is entirely visible. The Xerox

Star, released in 1981, was the first influential machine conceived as a workplace; its developers wrote:

> A well-designed system makes everything relevant to a task visible on the screen. . . . A subtle thing happens when everything is visible: *the display becomes reality.* The user model becomes identical with what is on the screen. Objects can be understood purely in terms of their visible characteristics. Actions can be understood in terms of their effects on the screen. This lets users *conduct experiments* to test, verify and expand their understanding— the essence of experimental science.[62]

Alan Kay summarizes the manipulatory externalized intellectual activity of the creative user: "Doing with images makes symbols."[63]

Language has lost its symbolic tongue, being replaced by the manipulation of pictures. The screen is now full of symbols in the form of icons, which can be handled by "clicks" or "drag and drop." "Users of this type of system have the feeling that they are operating upon the data directly, rather than through an agent."[64] Manipulation is ruled by the regime of the gaze. "What You See Is What You Get" is its motto, and nothing should happen behind the user's back. The experimental activity in the interface runs completely contrary to the top-down conception of structural programming. "Almost everything to the iconic mentality is 'before-after,' like a bird building a nest. The current state of things suggests what to do next. Extensive top-down planning is not required—just squish things around until you like the total effect."[65] While structured programming aimed at industrial production, now there are (inter)active individuals in their studios, handling icons without a predetermined plan. They have lost language and bricolage with information. "People who *do things*, like to have available a standard set of 'building blocks' whose properties they understand, with an 'escape' to new tool building when needed."[66] Alan Kay even imagines this speechless acting in the context of home: "The key is to find a context in which most of the things they want to do are as obvious as, say, moving furniture around in a house."[67]

Meanwhile, it had become obvious that software could not be produced in a factory and that problems of communication during the development process could not be solved by top-down organization. Software engineers also thought increasingly in such terms of building, which had more to do with Alan Kay's nestbuilding activity than with hierarchically structured production. Beginning in the 1980s, in the next and current phase of object-oriented design and object-oriented programming languages, the metaphor of *growing* software replaced the structural paradigm of rational construction in a closed system. This shift reflects the insight that the software production process has its own logic that cannot be predetermined by the structure of the anticipated product. "Program development as a social activity"—the title of an article by Kristen Nygaard—is now considered an evolutionary activity that reflects his slogan: "To program is to understand."[68] With "programming in the large," it is not possible to begin with the "metaphorical" operation of completely specifying a closed problem-space, then to deduce from such a "frozen specification" a blueprint and the synthesis of a program according to fixed rules. Programming is in itself an essentially "metonymical" process of knowledge acquisition.[69]

This means that it is no longer possible to maintain the separation of the different phases during software development. It is necessary to integrate the processes of clarification and communication in programming.[70] In practice analysis, design and programming are necessarily intertwined, and therefore a programming language has to provide means to fix intermediate results of a gradually growing experience with the problem domain. Object-oriented program systems can be grown. They have no predetermined structure, but rather are composites of model fragments, each of which represents a known and understood part of the world to be modeled and can be refined and modified. Their cohesion follows from a message-passing mechanism resembling the communication among their independently working developers. With this approach, language returned into the programming world, and computer scientists saw new facets of it. Language is no longer limited to the descriptive function of representing action in a formal model; rather, the discursive activity of modeling itself is somehow expressed in the medium of the programming language. If the distribution of production can no longer be derived from properties of an a priori specification of the problem, then it stands to reason that the experiences of the organized development process will find expression in the anthropomorphic artifacts of programming environments. Here, one finds objects that interact and *communicate*, exchange *messages*, are administered by *managers*, make use of services as *clients*, and keep *contracts*. One can say that the products in an object-oriented system with their interacting capabilities enact the process of their production. And as this process depends most importantly on communication, it means that "object-oriented programming relies on the concept of dealing with a *community* (a system) of *communicating objects*."[71]

In object-oriented programming languages, the character of program texts has changed. Object-oriented ensembles function similarly to a hypertext, whose stories are composed only by reading. They no longer represent the "narrative structure" of computation, but resemble more an aggregation of fragmentary plots; "there is no nesting of program texts. Systems are built as combinations of autonomous software elements"—the so-called classes.[72] These represent model fragments and are produced in stock for reuse, entailing an economical production of reusable texts. The boundaries of programming languages have become permeable. They are embedded in programming environments that supply, among other things, the model fragments in class-libraries. These part stores for modeling unite the collection of self-contained program texts in the old subroutine-libraries with the bottom-up building paradigm. Taking a long-term view, classes serve as ready-made building blocks for convenient use in the montage of new executable text collections, still called programs. The discipline of software engineering has finally come to an entirely pragmatic view of programming.[73]

The insight that programming is a social activity, which cannot be planned and organized in a centralized manner, and that design—as a process of growing understanding—needs to proceed incrementally, dealing with model fragments in a flexible way is reinforced in the newest approach of agent-oriented programming, beginning in the 1990s. Software-agents are objects that can act by themselves, or as it is said more pretentiously: they act "autonomously." They interact and cooperate with each other and sometimes also with their users. Agent-oriented design creates open, distributed systems with the help of autonomous modules—the agents. This requires abandoning control of the whole and passing it over to its components. This is true in design, which excludes a central perspective, as well as in the later behavior of the whole system, in which there is no unifying authority. "A fully decentralized approach will be essential. Rather than imposing an organizational structure from outside, no matter how extensible and flexible it might be, the system needs to be dynamically self-organizing."[74] Insofar as no controlling authority exists, either in the design of an open distributed system, or in the interplay of its autonomous components (the behavior of which may be moreover triggered interactively by human "components" in an unforeseeable way), the human as well as the artificial actors have no choice but to coordinate their performances. They find themselves acting out a grand collaboration without a director.[75]

Such an approach turns the top-down doctrine of structured programming (with its rationally controlled construction from a frozen specification of the whole), upside down and goes beyond the bottom-up procedure of object-orientation insofar as a design with autonomous modules exhibits greater interest in their interactions than their actions. In this view, correct behavior of the whole system results as an emergent property of these interactions. Kearney writes:

> In my view, a key goal in studying agents is to learn how to build large-scale systems with beneficial emergent behavior. By this I mean that individual agents are designed independently with behaviors that guide them in pursuit of their own goals and in their interactions with other agents. The properties of the system as a whole are not dictated by the designer, but arise as a result of repeated interactions.[76]

One might call modeling with agents a postmodern approach, insofar as it renounces control and welcomes contingency, puts confidence in the world, and "suggests that the world's independence of your control is not an obstacle to be overcome but a resource to be made use of."[77] Humboldt's conception of language and *Geist* is certainly under attack by object-oriented, and even more by agent-oriented, design and its languages. In distributed systems, the grand narrative of the algorithm has disappeared, with the loss of the subject that makes the design coherent. Since the self-aware *Geist* has given up, the "ordering of thought" can only be expressed in fragments.

Agents also invade the interface and are at the root of the newest interactive model. Rather than pottering about the work surface of an interface, the user now meets the computer in an "interspace." In this interworld, the performance of the computer and human activity are represented uniformly as both communicative and performative (inter)action. Agents of different kinds, both human and artificial, all live, act, interact, communicate, and cooperate in these worlds on an equal basis. This new equality requires that the users no longer perceive the computer as a supply of dead tools, but rather as an environment of "living" artifacts. Brenda Laurel in her book *Computers as Theatre* suggested conceiving the interface as a stage, in which human and machine agents perform together.[78] Roles can be attributed to the software agents, which indicate their performance. As they "act on behalf of the user," the user can delegate tasks to them and expect service, according to their role.

When we delegate tasks to a software agent, such as a search engine, we take on a different attitude toward artifacts, as if using a tool to intervene with in the world. We are used to delegating tasks to other human beings. Alan Kay refers to the social grip on others with the term "management": "We manipulate tools but manage people."[79] The transference of this idea of management to software agents thus signals deep changes in our attitude toward the computer. We now tend to anthropomorphize artifacts that can be delegated tasks, independently of their outer appearance, because we have socially influenced expectations of these artifacts and judge their behavior according to their willingness and their capacities. By delegating tasks to agents we lose oversight and control of their execution. Agents are not handy tools, which we can handle more or less skillfully; instead we have to ensure of their performance if we want to use them. We have left the "magical world of controllable media" and depend on a sympathetic worldview, when we evaluate the opaque contributions of agents according to our social experiences.[80] Thus, the view of the computer as a tool has given way to an interactive medium that (again) exhibits social characteristics.

Computer scientists also speak of interaction with agents as an "indirect manipulation": "Instead of user-initiated interaction via commands and/or direct manipulation, the user is engaged in a cooperative process in which human and computer agents both initiate communication, monitor events and perform tasks."[81] Cooperation between human and machine only works when the user can easily make his or her wishes to the agent clear. Unlike Ross's Gestalt-language, which concentrated on the exchange of *ideas*, the emphasis is now on the communication of *goals*. Corresponding to the user's distinction between direct and indirect manipulation is the machine's contradistinction between explicit and implicit responsiveness. There is no advantage to the user's needing to enter his or her intentions with explicit commands, following the logic of the machine. Rather, the agent should be able to "understand" the intentions and preferences of the user and to transform them into successful actions. "Whether those goals are explicitly stated by the user or inferred by the system, the way an agent interprets and attempts to meet them constitutes *implicit responsiveness*. This is the principal means whereby an agent amplifies the user's personal power."[82]

The principle of direct manipulation is turned upside down and the visible field is left behind when work is handled indirectly, processes are run behind the user's back, and artifacts react implicitly to the user's intentions. Relevant aspects of a task are now situated in the opaque realm of interpretation, cooperation, and negotiation. Often, the user cannot precisely specify the task to be delegated, or clearly recognize the competence of an agent. "What you see is what you get" does not apply to agents; rather, what one gets is often astonishing. The user cannot grasp the extent or quality of a service, but rather has to infer or to "sense" it. In order to make the user at home with the obscure behavior of artifacts, Brenda Laurel proposes an anthropomorphic representation of the agents themselves in the form of distinctive characters whose traits should help one to infer the performances that can be expected.[83]

Since delegation requires the investment of trust, independent of the agent's strength of character, it poses the issue of its trustworthiness. Character traits alone do not reveal the complex service offered. Rather the agent should be able to explain itself: "It will not be an agent's manipulative skills, or even its learning abilities, that will get it accepted, but instead, its safety and ability to explain itself in critical situations."[84] However, explanation alone will not suffice; in the end the conjunction of wishes and performance requires cooperation and negotiation. "We generally need to be assured that the agent shares our agenda and can carry out the task the way we want it done. This generally requires a *discourse* with the agent, a two-way feedback, in which both parties make their intentions and abilities known, and mutually agree on something resembling a *contract* about what is to be done, and by whom."[85] Thus a metadiscourse about the partner's possible contribution replaces the problem-oriented discourse between man and machine in the early interactions.

FUNCTION, REPRESENTATION, AGENCY OR: VERB, NOUN, AND DISCOURSE

Naturally, programming always has to be task- or process-oriented, because in the end, something is to be executed. However, there are significant variations in the modeling of a task, as well as in the representation of the part of the world that inspires this task. In the beginning, a programmable solution meant direct reference to computation. Accordingly, functional aspects were dominant, and *procedures* were the principal features of programming languages.[86] Data structures only indicated different modes of processing and their necessary storage space. The focus on functional or algorithmical aspects of computing had the effect of delaying consideration of the nature of abstraction in modeling and of developing adequate methods for abstracting concepts. Many people were not even conscious that it was not processing, but modeling that was the core issue.

One can describe this phenomenon as not using the language analogy to its full potential. The procedural programming languages organize "verbs" as dominating elements of language, because procedures describe activities as verbs do in natural language. The man-machine conversation was "verbally" oriented, too, because the focus of interest was on the interactive procedure. However, it is said that the primary operation in human use of language is naming and that activities are related to named objects. At least philosophers (in Western cultures) have always shown a predilection for nouns, which provided for them a more orderly and stable grasp of the world.[87] Morris Kline quotes Jeremy Bentham in his book *Mathematics in Western Culture*: "Nouns, he said, are better than verbs. An idea embodied in a noun is 'stationed on a rock'; one embodied in a verb 'slips through your fingers like an eel.' The ideal language would resemble algebra; ideas would be represented by symbols as numbers are represented by letters."[88] If one takes this view, then "natural" formal modeling should rely on data structures, which represent entities of reality, and all operations should be associated with the data objects for which they are defined. And similarly, as complex terms (modeled as compound nominal phrases) can be composed from simpler ones, and abstract terms can be extracted from an aggregation of similar more concrete ones, a programming language, aimed at grasping reality in a natural and "rock-solid" way should support the construction of formal imitations of familiar notions of the problem sphere at different levels of abstractions. It is the objective of object-oriented programming languages—further discussed below—to fulfill these two requirements for "natural" modeling.

The procedural view of language in computer science, however, clung to computation and was not able to formally describe the entities to be modeled. Imperative languages were conceived from the standpoint of an analyzing automaton—the well-known "mechanical reader"; therefore grammar could only play its synthesizing counterpart, constraining the generation of texts so that they could be easily "interpreted" by the automaton. Such a language concept could not serve as a frame for specifying adequate models. Names occurred only as arbitrary identifiers, symbolic addresses for memory cells in the machine; procedures were understood as little more than abbreviations for program pieces, which saved storage and writing labor, and not as independent conceptual units.

In opposition to the actual practice of programming, early on there was a desire for a language that would allow problems and relations between data to be stated without getting involved in the turmoil of computation. The model, again, was algebra. The Codasyl Language Structure Group remarked in 1962, "With current programming languages, the problem definition is buried in the rigid structure of an algorithmic statement of the solution, and such a statement cannot readily be manipulated." They proposed an "Information Algebra" as a solution, which would state relations between real entities (represented in the system as data) only with the help of operations based on these data. "The primary intent of the Information Algebra is to extent the concept of stating the relationships among data to all aspects of data processing."[89] The appeal came too early, and one cannot really speak of "natural modeling" in connection with algebra.

Nevertheless, realizing the importance of data structures led to a change in point of view from algorithmic procedures to the modeling of entities and their relations, although rather slowly.[90] I have mentioned earlier that Maurice Wilkes in 1967 still argued against the "linguistic turn" in computer science, which meant at that time preoccupation with grammar and compiler construction, for the sake of better performance. He justified his preference for data structures:

> At the present time, choosing a programming language is equivalent to
> choosing a data structure, and if that data structure does not fit the data you
> want to manipulate then it is too bad. It would, in a sense, be more logical
> first to choose a data structure appropriate to the problem and then look
> around for, or construct with a kit of tools provided, a language suitable for
> manipulating that data structure.[91]

Wilkes's criticism was aimed at the fact that every modeling of entities in the problem domain has to do essentially with data structures and that a procedure in which one chooses a fixed set of data types together with a programming language is an inflexible and inefficient one.

Probably the first glimpse of formal *representations* of "true notions" modeling real-world entities arose from a new feature in some programming languages: heterogeneous composite data structures definable by the programmer. These so-called records or structures allowed the combination of related elements of the modeled sphere and the treatment of them as an entity, without having to attend the machine's internal affairs. One can group things and properties according to logical viewpoints. This means that data are no longer grouped by similar data formats in the computer, but according to shared features in observed or constructed entities of reality. Since no specific activities could be assigned to these structures, they represented a rather static view of the world.

In structured programming one tried likewise to escape the machine's constraints and develop a more abstract conception of functionality. A concept of modularization was developed, in which the interface of a module should describe only *what* the module does and not *how* it is done. Thus, modules encapsulate pieces of functionality that belong together, hiding as much information as possible about the actual processing. Even though the importance of data structures was well accepted in structured programming, the dominance of functions remained unaffected. The top-down design of a program is based on the decomposition of functionality into subfunctions. Similarly, other developments in which programming proceeded more declaratively did not focus on technical equivalents of descriptive notions; they oriented themselves by algebras or operated with predicates as in logic programming.

Meanwhile, it had been realized—fully in accordance with Jeremy Bentham—that in the process of software development it is functionality that changes most with understanding and growing experience, whereas data structures, modeling observable properties or notions of the problem domain, were the more stable parts. Now, data structures were perceived as being important, not, as Wilkes did, because of considerations about efficiency, but because of their capability to reliably describe persistent parts in the modeling process. The consequence of object-oriented programming was that its fundamental elements—the *objects*—are based on data structures, to which operations are assigned. Objects exist in fact only during runtime. They are created as instances of an abstract description—of their *class*. A class in object-oriented programming languages is a module in which the relevant properties of a class of similar entities of reality and what can be done with them are defined.[92] The properties are represented by data structures and the class is equipped with *methods* that are actually nothing but procedures or functions. These can be invoked by *messages* from other objects but can operate only on the objects to which they belong. Thus, properties take precedence over activities related to them.

I would like to show the difference between procedural and object-oriented modeling with an example slightly outside computer science. In order to express that Socrates died, the functional approach would define the procedure "dying" and apply it on the datum "Socrates" in order to obtain the desired effect. In an object-oriented language one would define the class "human," which contains the Boolean property "living/dead" and the method "dying." After the creation of an object-instance (belonging to the class "human") with the name "Socrates," it can be subjected to its operation "dying" and takes on the property "dead." Thus, the object-oriented language inventory contains model fragments that indicate the state of an entity ("noun") and describe how this state can be changed ("verb").[93]

Classes are formed in analogy with conventional notions. This formal equivalence is strengthened by way of a mechanism of inheritance. One can refine classes into *subclasses*. These inherit the properties and methods of their *superclass*, but they can also be extended or modified. (Thus, one would probably have defined in the example above a subclass "philosopher" and modified the method "dying" with respect to suicide out of principle considerations.) Inversely, one can abstract a superclass out of various similar (sub)classes containing its common parts. An object-oriented programming system thus represents the operational form of a hierarchy of concepts in classical logic, which is structured by the order relation of the logical categories *genus* and *species*. Further, the time-honored terms of philosophical logic *extension* and *intension* of a concept are represented by object-instances and class-definitions. The claim is that in such a way it is possible to model in a more natural way, because the programmer can more straightforwardly use preexisting notions (even use their familiar names); furthermore, such a technique would imitate the "natural" human approach to the world, which labels objects first and does something with them second.

Objects, like tools, are inherently passive and need activation from outside. However, in the spaces of the Net and with multimedia applications one needs objects that are always active and act "autonomously." These take on the form of *agents* and can be delegated tasks. Their capacity to act independently on behalf of the user, is not a regression to the early motive of replacing the human as in automation; their acting as representatives is meant for service. Agents resemble robots more than a fully automated factory; thus they are also called "softbots." They act most often in agent societies.

The technical realization of "autonomy" can be tackled with very different agent architectures. Models that imitate the human intellect are widespread. They are provided with so-called mental states, which represent knowledge, assumptions, capacities, values, commitments, aims, and intentions; one attributes "beliefs, desires and intentions" to these agents (in so-called *BDI*-architectures).

> By taking action, agents actively attempt to satisfy a value system that describes what is desirable. In order to satisfy their values, agents derive goals from them and then form intentions to take actions to reach these goals. . . . Agents know facts about their environment and about themselves. They communicate with each other by sending messages containing declarative facts, requests for action, declaration of values or goals.[94]

Whereas passive objects have no control over their actions, when triggered by a message from other objects, an agent that receives a request can deny it according to its inner "value-system," or do something else. "Because of this distinction, we do not think of agents as invoking methods (actions) on agents —rather, we tend to think of them *requesting* actions to be performed."[95] Agents do not only provide, like objects, a data abstraction by hiding implementation specifics; they also encapsulate "behavior," meaning they offer a *service*. Whereas an ensemble of objects (more precisely their classes) represents part of the world rather statically in the form of a hierarchy of concepts, agents act according to *roles* and are organized in a society of agents by this social characteristic.[96] The roles indicate *what* can be expected of them and hide *how* they fulfill these expectations. Modeling of and with agency thus represents a powerful form of task abstractions.

One can consider this refined technique of "information hiding" as an increased attempt to model formal objectifications—artifacts—to conform more directly to their mental origin—"mentefacts." The "mode of connection of ideas" forced by agent-oriented modeling should make it possible, to recognize better the intended "mentefacts" in the enacting of the executable artifacts. Agent-based languages thus represent a new medium of reflection on design. Whereas the first problem-oriented languages were a means of communicating about algorithms, and object-oriented models are to facilitate the representation of operational entities of the problem sphere, agents can help developers to clarify their own beliefs, desires, and intentions: "Agents represent an opportunity for product designers to rethink some of the underlying premises of how they conceive of functionality."[97]

At the same time this new design technique stands in the tradition of software engineering. It is claimed that only now can the two fundamental measures to reduce complexity in software development—*modularization* and *abstraction*—be realized in their full sense. The composition of an open system from independent modules whose interactions constitute its global behavior, (or the corresponding decomposition into modules) often brings to bear a better modularization than a central perspective. "This decomposition allows each agent to employ the most appropriate paradigm for solving its particular problem, rather than being forced to adopt a common uniform approach that represents a compromise for the entire system, but which is not optimal for any of its subparts." By providing a forceful task abstraction, agents "allow a software developer to conceptualize a complex software system as a society of cooperating autonomous problem solvers. For many applications, this high-level approach is simply more appropriate than the alternatives."[98]

It is said that designing systems with autonomous modules suffices to make the artifacts show a behavior similar to human beings and the interaction between the unequal partners look natural: "Autonomous and modularized mean having human-like behavior. . . . A primary characteristic of agent-based systems is that the collaboration among human and artificial agents is more natural and seamless."[99] Agency concerns developers and users in a similar way, thus emphasizing the increasing affinity of their activities. Agents imitate the behavior of developers and users, who meet in the expanse of the information nets and cooperate with each other. Agents need not only to be adapted in the interface to their user, but they should behave among each other like their human models. "The agent metaphor comes into its own when we provide means of interaction between independent agents, so that the agents themselves form a distributed information system. . . . So the image of an agent-based distributed information system is of a community of peers that provide services to each other on terms which are negotiated to the mutual satisfaction of both agents."[100]

The agent introduces moments of social communication in the programming world; interaction, cooperation, and negotiation turn into the fundamental modes of operation of the modules. At the heart of this conception lies therefore a broader understanding of language that includes discursive and pragmatic elements. Its artifacts behave like performing "speech acts" in such a way that they incarnate the cognitive and communicative activities of their developers and users. If we model the dying Socrates as an agent, it could, according to its role as a philosopher and according to its inner "value-system," decide not to drink the hemlock, negotiate instead with the rulers (modeled as a different kind of agents), and accommodate itself to them. I am not claiming that one could not conceive of modeling such a complex behavior before. We could have programmed the "dying"-procedure with all kinds of possibilities, or we could have designed the object-oriented method "dying" in such a way that the philosopher remains alive under certain circumstances. It is decisive that in agent-modeling we do not plan the context of such a "decision" for the whole system, but that it follows in the context of a "formal language-game" from the interactions and negotiations of the individual actors.[101]

CONCLUSION

It seems to me that the common history of the three outlined perspectives is that in each the original aspects of language retreat to the background and aspects of discourse have taken the stage. The focus of attention now lies on communication and cooperation, on the arrangement of different contexts in which these can take place, and on the representation of performative interactivity. Interest in the executive function of formal languages made room for the pragmatic question of how to represent communicative action.[102] In the sense of the theory of speech acts, the nondescriptive moments of a speech act are expressed in the medium of a formal language and incorporated in the artifacts. In this view, the different perspectives of modeling appear to be represented more congruently, insofar as such artifacts more directly reflect cognitive acts, communicative activity, self-communication, and disposition of developers and users. Objective representation of knowledge in the form of interacting agents is both a condition and a representation of the subjective process in which this knowledge is created and dealt with. In modeling with agency, the externalization of subjectivity, inherent in the interplay of "inner and outer discourse," is technically "incarnated" in autonomous agents insofar as their interactions objectify fragmentary and contingent processes of communication and cognition.

One might say that the languages of computer science have become more "subjective" over time, which explains perhaps that the formerly busy research agenda of "theory of formal languages" has become rather insignificant. According to the dialectics a language "is objectively active and independent, precisely insofar as it is subjectively passive and dependent."[103] This also means that subjectivity becomes more formal through subjects communicating with the artifacts. Computer science as a social technology, too, solves its problems of modeling to a large extent by proceeding according to Douglas Ross's rule that part of the problem can be solved by "programming the human."[104] When the interesting potential of the computer has become its "capacity to *represent action in which humans could participate*," then the performance of the user acting in a virtual *theatrum mundi* is the program.[105] In the performance in which both human and artificial agent act and interact, represent themselves and cooperate, their roles become more and more entwined and interdependent.[106] The prophets of Artificial Intelligence call on us to be prepared: "We have to think here's one intelligent agent, there's another intelligent agent and they have complementary capabilities. We have to design our systems so that both of these are working together to produce a better result."[107]

I believe that the *Geist*, in submitting to this cooperation, indicates a new "order of knowledge," an epistemological change in dealing with knowledge not limited to direct relationship with the computer. The winding paths of language and intellectual activity, which Humboldt regarded as having a common origin in the essence of the human being, ends up in the technically mediated formality of interactivity. The inwardness of thinking, once understood as a "tentative action" in the form of an inner discourse, is pushed aside by an energetic interactive dealing with knowledge in information spaces. The appropriate behavior of the user in "the vast, rich, ever-changing world of electronic communication and information" can be defined as acting with information mediated by artifacts and a common formal language.[108] The concept of *interactivity*, derived from the coupling of human and machine, thus manifests itself as a leitmotif of the evolving information society.

REFERENCE MATTER

Chapter 1

This chapter aims to give a historical overview of analogies to language in the sciences, building on the work contained in this collection. I would like to thank Cathryn Carson, Lorraine Daston, Abigail Lustig, and Dorinda Outram for their indispensible commentaries and suggestions.

1. Peter Dear, ed., *The Literary Structure of Scientific Argument* (Philadelphia: University of Pennsylvania Press, 1991), 4–5.

2. Just to name some prominent representatives: Charles Bazerman, *Shaping Written Knowledge* (Madison: University of Wisconsin Press, 1988); Alan G. Gross, *The Rhetoric of Science* (Cambridge, Mass.: Harvard University Press, 1990); Greg Myers, *Writing Biology: Texts in the Social Construction of Scientific Knowledge* (Madison: University of Wisconsin Press, 1990).

3. Marco Beretta, "The Grammar of Matter: Chemical Nomenclature During the XVIII Century," in *Science et Langues en Europe*, ed. Roger Chartier and Pietro Corsi (Paris: EHESS, 1996), 109–25, 110.

4. " . . . dominés et transis par le langage," in Michel Foucault, *Les mots et les choses* (Paris: Gallimard, 1966), 311; Michel Foucault, *The Order of Things* (London: Routledge, 1970), 298.

5. See Jeanne Fahnestock, *Rhetorical Figures in Science* (Oxford: Oxford University Press, 1999), xii. Fahnestock wishes to revive historical and particularly Aristotelian rhetoric, pointing to a whole range of other figures of speech beside metaphors that function as epitomes.

6. This book is not concerned with the unmediated generation of new thoughts during the act of speaking as Heinrich v. Kleist argued in his famous letter *Über die allmähliche Verfertigung der Gedanken beim Sprechen* (1805/1806).

7. Max Black, *Models and Metaphors: Studies in Language and Philosophy* (Ithaca, N.Y.: Cornell University Press, 1962).

8. Stephen G. Alter, *Darwinism and the Linguistic Image* (Baltimore, Md.: Johns Hopkins University Press, 1999); Robert Brain, "Standards and Semiotics," in *Inscribing Science: Scientific Texts and the Materiality of Communication*, ed. Timothy

Lenoir (Stanford, Calif.: Stanford University Press, 1998), 249–84; Matthias Dörries, "Heinrich Kayser as Philologist of Physics," *Historical Studies in the Physical and Biological Sciences* 26, no. 1 (1995): 1–33; Brigitte Nerlich and David C. Clarke, "The Linguistic Repudiation of Wundt," *History of Psychology* 1, no. 3 (1998): 179–204.

9. Max Planck, "Positivismus und reale Außenwelt. Vortrag 1930," in *Wege zur physikalischen Erkenntnis I*, 2d ed. (Leipzig: S. Hirzel, 1934 [1933]), 208–32, 218.

10. See Gillian Beer, *Darwin's Plots: Evolutionary Narrative in Darwin, George Eliot, and Nineteenth-Century Fiction* (London: Routledge, 1983).

11. Alexander von Humboldt, *Cosmos*, 5 vols., trans. E. C. Otté (London: Henry G. Bohn, 1848–58), 1:37. " . . . so ergießt sie [die Sprache] zugleich . . . ihren belebenden Hauch auf die Gedankenfülle selbst" (A. v. Humboldt, *Kosmos*, 5 vols. [Stuttgart: Cotta, 1845–62], 1:40).

12. See Owen Hannaway, *The Chemists and the Word: The Didactic Origins of Modern Chemistry* (Baltimore, Md.: Johns Hopkins University Press, 1975).

13. Hans Blumenberg, *Die Lesbarkeit der Welt*, 2d ed. (Frankfurt am Main: Suhrkamp, 1983). Heribert M. Nobis, "Buch der Natur," in *Historisches Wörterbuch der Philosophie*, 10 vols., ed. Joachim Ritter (Basel: Schwabe, 1971–98), 1:957–59. James J. Bono, *The Word of God and the Languages of Man: Interpreting Nature in Early Modern Science and Medecine*, vol. 1 (Madison: University of Wisconsin Press, 1995).

14. Stillman Drake and C. D. O'Malley, *The Controversy on the Comets of 1618* (Philadelphia: University of Pennsylvania Press, 1960), 183–84. "La filosofia è scritta in questo grandissimo libro che continuamente ci sta aperto innanzi a gli occhi (io dico l'universo), ma non si può intendere se prima non s'impara a intender la lingua, e conoscer i caratteri, ne'quali è scritto. Egli è scritto in lingua matematica, e i caratteri son triangoli, cerchi, ed altre figure geometriche, senza i quali mezi è impossibile a intenderne umanamente parola; senza questi è un aggirarsi vanamente per un oscuro laberinto" (Galileo Galilei, *Il saggiatore*, in *Opere*, ed. Antonio Favaro, 20 vols. [Florence: Barbera, 1890–1909], 6:232).

15. See the *Liddell-Scott Lexicon of Classical Greek*.

16. See Geneva's chapter in this book.

17. M. M. Slaughter, *Universal Languages and Scientific Taxonomy in the Seventeenth Century* (Cambridge: Cambridge University Press, 1982), vii–viii.

18. It might also be interesting to explore how far written texts are results of forms of rhetoric and verbal communication. Often enough the texts were born in dictation to a writer, which explains redundancies and other rhetorical forms.

19. François Jacob et al., "Vivre et parler: Un débat entre François Jacob, Roman Jakobson, Claude Lévi-Strauss et Philipp L'Héritier," part 1, *Les lettres françaises* 1221 (February 14, 1968): 3–7, 6. For a recent book on genetics and language, see Lily E. Kay, *Who Wrote the Book of Life? A History of the Genetic Code* (Stanford, Calif.: Stanford University Press, 2000).

20. Language and combinatorics issues might be extended to chemistry (as a

combinatorial art), as suggested by Pierre Laszlo in his *La Parole des Choses ou le Langage de la Chimie* (Paris: Hermann, 1993).

21. Marco Beretta, *The Enlightenment of Matter: The Definition of Chemistry from Agricola to Lavoisier* (Canton: Science History Publications, 1993), chap. 3.

22. Zdravko Radman, *Metaphors: Figures of the Mind* (Boston: Kluwer Academic Publishers, 1997), 56–57.

23. See Geneva's chapter in this volume.

24. Beretta, *Grammar of Matter* (note 21).

25. Jessica Riskin, "Rival Idioms for a Revolutionized Science and a Republican Citizenry," *Isis* 89, no. 2 (1998): 203–32, 205.

26. Ibid., 216 (note 25).

27. Peter F. Stevens, "Metaphors and Typology in the Development of Botanical Systematics 1690–1960, or the Art of Putting New Wine in Old Bottles," *Taxon* 33 (1984): S.169–211. Peter F. Stevens, *The Development of Biological Systematics: Antoine-Laurent de Jussieu, Natur, and the Natural System* (New York: Columbia University Press, 1994).

28. Etienne Bonnet de Condillac, *The Logic of Condillac*, ed. Daniel N. Robinson (Washington, D.C.: University Publications of America, 1977), 66. See also Robin E. Rider, "Measure of Ideas, Rule of Language: Mathematics and Language in the 18th Century," in *The Quantifying Spirit in the Eighteenth Century*, ed. Frängsmyr, Tore et al. (Berkeley and Los Angeles: University of California Press, 1990), 113–40.

29. Lavoisier, Guyton de Morveau, Fourcroy, Berthollet, *Method of Chymical Nomenclature* (London, 1788), 4–5. "Les langues n'ont pas seulement pour objet . . . d'exprimer par des signes, des idées & des images: ce sont de plus de véritables méthodes analytiques, à l'aide desquelles nous procédons du connu à l'inconnu, & jusqu'à un certain point à la manière des mathématiciens" (Lavoisier, Guyton de Morveau, Fourcroy, Berthollet, *Méthode de nomenclature chimique* [Paris, 1787], 6).

30. Beretta, *Grammar of Matter*, 125 (note 21).

31. Herbert Mehrtens, *Moderne Sprache Mathematik: eine Geschichte des Streits um die Grundlagen der Disziplin und des Subjekts formaler Systeme* (Frankfurt am Main: Suhrkamp, 1990). Eberhard Knobloch, "Analogie und mathematisches Denken," *Berichte zur Wissenschaftsgeschichte* 12, no. 1 (1989): 35–47.

32. This term is from Radman, *Metaphors, Figures of the Mind*, 45 (note 22). The following paragraph is indebted to his chapter 4, "The Metaphoric Measure of Meaning in Science."

33. E. R. MacCormac, *Metaphor and Myth in Science and Religion* (Durham, N.C.: Duke University Press, 1976), 136.

34. James C. Maxwell, "Address to the Mathematical and Physical Section of the British Association (15 Sept. 1870)," in *Scientific Papers*, 2 vols., ed. W. D. Niven (Cambridge: Cambridge University Press, 1890), 2:215–29, 227.

35. Werner Heisenberg, "Prinzipielle Fragen der modernen Physik," in *Neuere Fortschritte in den exakten Wissenschaften (Fünf Wiener Vorträge, 3. Zyklus)* (Leipzig/ Wien: Franz Deuticke, 1936), 91–102, 93–94.

36. Heisenberg saw similar problems with logical positivism; see the quotation on page 15.

37. Erich Rothacker, *Das "Buch der Natur". Materialien und Grundsätzliches zur Metapherngeschichte* (Bonn: Bouvier, 1979), 13.

38. The study of language has been central to such different cultures as Europe, India, and China. This would lead to another project about how these very different languages informed and inform views on nature and the world, part of a *longue histoire des textes* of the sort suggested by Karine Chemla in "Histoire des sciences et materialité des textes," *Enquête* 1 (1995): 167–80; see also for an English version, Karine Chemla, "What Is the Content of this Book? A Plea for Developing History of Science and History of Text Conjointly," *Philosophy and the History of Science* 4, no. 2 (1995): 1–46.

Chapter 2

Cheryce Kramer made many suggestions for the improvement of this essay, and Paul White furnished essential bibliographic help. I am grateful to both for their generous assistance.

1. For a brief discussion of Darwin's later views about language, see Robert J. Richards, *Darwin and the Emergence of Evolutionary Theories of Mind and Behavior* (Chicago: University of Chicago Press, 1987), 200–206.

2. Charles Darwin, *Old and Useless Notes* in *Charles Darwin's Notebooks, 1836–1844*, ed. P. Barrett et al. (Ithaca, N.Y.: Cornell University Press, 1987), 599 (5–5v).

3. See Charles Darwin, *Notebook N*, in ibid., 568, 581 (18 and 65) (note 2).

4. See Hensleigh Wedgwood, *On the Origin of Language* (London: Trübner, 1866), 13–14, 129. See also Richards, *Darwin and the Emergence of Evolutionary Theories*, 205 (note 1).

5. In actual bulk of pages, Darwin's two-volume *Descent of Man and Selection in Relation to Sex* swims in data and discussions of sexual selection in butterflies, birds, and other beasts. The aim of this investigation, though, is to shed light on human traits and human sexual dimorphism. Charles Darwin, *On the Descent of Man and Selection in Relation to Sex*, 2 vols. (London: Murray, 1871).

6. I have discussed Wallace's spiritualistic interpretation of evolution and Darwin's reaction in *Darwin and the Emergence of Evolutionary Theories of Mind and Behavior*, 176–84 (note 1).

7. Darwin, *Descent of Man*, 1:57 (note 5).

8. F. H. Bradley to C. Lloyd Morgan (February 16, 1895), in the Papers of C. Lloyd Morgan, DM 612, Bristol University Library.

9. Darwin, *Descent of Man*, 1:58, 2:390–91 (note 5).

10. John Locke (1632–1704), as usual, established the common British view. He held that God furnished man with language in order "to use these sounds as signs of internal conceptions; and to make them stand as marks for the ideas within his own

mind, whereby they might be made known to others, and the thoughts of men's minds be conveyed from one to another." Although thought used language, according to Locke, "thought is not constituted by, nor identical with language, which on the contrary is originated and formed by thought." See John Locke, *An Essay Concerning Human Understanding*, 2 vols. (New York: Dover, 1959 [1670]), 2:3, n. 2.

11. Charles Darwin, *On the Origin of Species* (London: Murray, 1859), 488.

12. For an analysis of the several places in the *Origin* where Darwin makes fleeting reference to human beings, see Kathy J. Cooke, "Darwin on Man in the *Origin of Species*," *Journal of the History of Biology* 26 (1990): 517–21.

13. Darwin, *Origin of Species*, 422 (note 11).

14. Charles Lyell, *The Geological Evidences of the Antiquity of Man* (London: Murray, 1863), 463.

15. Lyell, *Geological Evidences*, 469 (note 14).

16. Alexander von Humboldt not only conveyed a conception of living nature that Darwin incorporated into his own evolutionary theory, but he also suggested that language helped to create human intellect. In the English translation of Humboldt's *Kosmos*, which Darwin read in the 1850s, the following may be found: "But thought and language have ever been most intimately allied. If language, by its originality of structure and its native richness, can, in its delineations, interpret thought with grace and clearness, and if, by its happy flexibility, it can paint with vivid truthfulness the objects of the external world, it reacts at the same time upon thought, and animates it, as it were, with the breath of life. It is this mutual reaction which makes words more than mere signs and forms of thought; and the beneficent influences of a language is more strikingly manifested on its native soil, where it has sprung spontaneously from the minds of the people, whose character it embodies." See Alexander von Humboldt, *Cosmos*, 5 vols., trans. E. C. Otté (New York: Harper & Brothers, 1848–68), 1:56. This general view of language is also to be found in August Schleicher, as I explain below in the text. Both theorists, however, seem to have had a common source: Wilhelm von Humboldt (see below). For a discussion of the impact of Alexander von Humboldt's ideas on Darwin's conception of nature, see Robert J. Richards, "Darwin's Romantic Biology, the Foundation of his Evolutionary Ethics," in *Biology and the Foundations of Ethics*, ed. Jane Maienschein and Michael Ruse (Cambridge: Cambridge University Press, 1999), 113–53.

17. See Charles Darwin, *Über die Entstehung der Arten im Thier- und Pflanzen-Reich durch natürliche Züchtung oder Erhaltung der vervollkommenten Rassen im Kampfe um's Daseyn*, 2d ed. (from the 3d English ed.), trans. H. G. Bronn (Stuttgart: Schweizerbart'sche Verlagshandlung und Druckerei, 1863).

18. Schleicher was indeed a serious gardener and wrote a review of the *Origin* for an agricultural journal. See August Schleicher, "Die Darwin'sche Theorie und die Thier- und Pflanzenzucht," *Zeitschrift für deutsche Landwirthe* 15 (1864): 1–11. In the review, Schleicher summarized Darwin's argument and added elements that he undoubtedly thought rounded out the theory, including the suggestion that human beings descended from the "higher apes" and differed from them only by reason of

language and "high brain development" (6). Schleicher neglected to mention that Darwin himself did not discuss human evolution in the *Origin*. Schleicher sent this review to Darwin, and it is now held in the Manuscript Room of Cambridge University Library. Scorings indicated Darwin read the review.

19. August Schleicher, *Die Darwinsche Theorie und die Sprachwissenschaft* (Weimar: Böhlau, 1863). Recently two works have discussed Schleicher's book with skill and insight. See Liba Taub, "Evolutionary Ideas and 'Empirical' Methods: The Analogy Between Language and Species in Works by Lyell and Schleicher," *British Journal for the History of Science* 26 (1993): 171–93; and Stephen Alter, *Darwinism and the Linguistic Image* (Baltimore, Md.: Johns Hopkins University Press, 1999), esp. 73–79.

20. In the English press, Schleicher's book, in the first German edition, received immediate notice through an anonymous author. See "The Darwinian Theory in Philology," *The Reader* 3 (1864): 261–62. The author agreed with Schleicher that linguistics lent support to Darwin's theory. Friedrich Max Müller discussed the English translation (see next note) of the work in a review in *Nature* ("The Science of Language," *Nature* 1 [1870]: 256–59). Müller took exception to the idea that a descendent language sprang from a well-formed classical language (for example, French from Latin). Rather, he maintained that the descendent language arose from rude dialects that might trace their origin to the classical language. Frederick Farrar, feeling that Müller gave scant account of Schleicher's little book, provided a summary in a subsequent issue of *Nature* ("Philology and Darwinism," *Nature* 1 [1870]: 527–29). Darwin undoubtedly read these reviews. See the discussion of the controversy between Müller and Farrar in Alter, *Darwinism and the Linguistic Image*, 84–96 (note 19). William Dwight Whitney took grave exception to Schleicher's naturalism—that is, the supposition that languages displayed organic features and obeyed natural laws —and denied that Schleicher's notion of language descent gave any aid to Darwin's theory. See William Dwight Whitney, "Schleicher and the Physical Theory of Language" (1871), reprinted in his *Oriental and Linguistic Studies*, 2 vols. (New York: Charles Scribner's Sons, 1873), 1:298–331. Hans Arsleff details other responses to Schleicher's *Darwinsche Theorie*, of whose doctrines he thoroughly disapproves. See Hans Arsleff, *From Locke to Saussure: Essays on the Study of Language and Intellectual History* (Minneapolis: University of Minnesota Press, 1982), 293–334.

21. Darwin, *Descent of Man*, 1:56 (note 5). Darwin received a copy of *Darwinsche Theorie* (now held in the Manuscript Room of Cambridge University Library) from the author; in the *Descent* he referred to the English edition (also in the Manuscript Room of Cambridge University Library). See August Schleicher, *Darwinism Tested by the Science of Language*, trans. Alex Bikkers (London: John Camden Hotten, 1869). Marginal scorings indicate that Darwin read both versions of the book.

22. Heinrich Bronn, "Schlusswort des Übersetzers," in Charles Darwin, *Über die Entstehung der Arten*, 525–51. Bronn brought as a chief objection to Darwin's theory that it was "in its ground-conditions of justification still a thoroughly wanting hypothesis." It remained, according to Bronn "undemonstrated," although also "un-

refuted" (532–33). Bronn did, however, lodge some considerations that militated against the hypothesis, for example, that transitional species were lacking (534–35). Bronn himself was the author of a quasi-evolutionary theory, which he formulated prior to reading Darwin. He elaborated his theory in a prizewinning essay, selections of which were translated into English as "On the Laws of Evolution of the Organic World during the Formation of the Crust of the Earth," *The Annals and Magazine of Natural History*, 3d ser., 4 (1859): 81–90, 175–84. Bronn argued for a gradual appearance of new species and an extinguishing of more primitive ones over great periods of time. Such evolution did not involve, however, the transformation of one species into another, merely the successive appearance and adaptation of progressively higher kinds of flora and fauna. This process occurred, he strongly implied but did not expressly say, through Divine Wisdom. His views were not unlike those of Louis Agassiz and Richard Owen. For a discussion of the ideas of these latter thinkers, see Robert J. Richards, *The Meaning of Evolution* (Chicago: University of Chicago Press, 1992), 116–21.

23. Schleicher, *Darwinsche Theorie*, 4–8, 23–24 (note 19).

24. Ibid., 8 (note 19).

25. This is one of the general themes of my forthcoming *The Tragic Sense of Life: The Battle over Revolutionary Thought in Germany*.

26. August Schleicher, *Über die Bedeutung der Sprache für die Naturgeschichte des Menschen* (Weimar: Böhlau, 1865), 16 and 18–19.

27. Ibid., 21 (note 26).

28. See the conclusion of this paper for a discussion of Darwin's knowledge of Schleicher's *Bedeutung der Sprache*.

29. For details of Schleicher's life I have relied on Johannes Schmidt, "Schleicher," *Allgemeine deutsche Biographie* 31 (1890): 402–15; Joachim Dietze, *August Schleicher als Slawist: Sein Leben und sein Werk in der Sicht der Indogermanistik* (Berlin: Adademie-Verlag, 1966); and Theodor Syllaba, *August Schleicher und Böhmen* (Prague: Karls-Universität, 1995).

30. Syllaba, *August Schleicher und Böhmen*, 18 (note 29).

31. Syllaba characterizes Schleicher's work as a correspondent and provides a list of the articles in his *August Schleicher und Böhmen*, 13–27 (note 29).

32. See Dietze, *August Schleicher als Slawist*, 16 (note 29).

33. Robert Boxberger, "Prager Erinnerungen aus Jena," quoted in Dietze, *August Schleicher als Slawist*, 45 (note 29).

34. August Schleicher, *Zur vergleichenden Sprachengeschichte* (Bonn: H. B. König, 1848).

35. In distinguishing these three forms of language, Schleicher was simply following the lead of Wilhelm von Humboldt, Franz Bopp, and ultimately August Wilhelm Schlegel. Schleicher was certainly familiar with the work of these near contemporary linguists. In his *Sprachengeschichte*, he cited Humboldt often enough, although not precisely on this distinction. See Wilhelm von Humboldt, *Über die Kawi-Sprache auf der Insel Java*, 3 vols. (Berlin: Königliche Akademie der Wissenschaften, 1836).

The introduction to this famous work on Javanese language made the threefold distinction pivotal (1:cxxxv–cxlviii). August Wilhelm Schlegel, who became professor of linguistics at Bonn, formulated the original distinction in his *Observations sur la langue et la littérature provençales* (Paris: Librarie grecque-latine-allemande, 1818), 14–16. Franz Bopp, whom Humboldt brought to Berlin as professor, canonized the distinction in his *Vergleichende Grammatik des Sanskrit, Zend, Griechischen, Lateinischen, Litthauischen, Gothischen und Deutschen* (Berlin: Königliche Akademie der Wissenschaften, 1833), 108–13.

36. Schleicher, *Zur vergleichenden Sprachengeschichte*, 8–11 (note 34).

37. Humboldt, for instance, liked to refer to the internal coherence of language by use of the term "the language-organism" [*Sprachorganismus*]. See Humboldt, *Über die Kawi-Sprache*, 1:cxxxv (note 35). Bopp likewise generously employed the organic metaphor; as he expressed it in his *Vergleichende Grammatik*, iii (note 35): "I intend in this book a comparative, comprehensive description of the organism of the languages mentioned in the title, an investigation of their physical and mechanical laws, and the origin of the forms indicating grammatical relationships." Humboldt and Bopp had, in utilizing this metaphor, adopted the conception of Friedrich Schelling, the philosophical architect of the romantic movement. See, for instance, a typical observation of Schelling, in his *Historisch-kritische Einleitung in die Philosophie der Mythologie* (1842), in *Friedrich Wilhelm Joseph von Schelling Ausgewählte Schriften*, ed. Manfred Frank, 6 vols. (Frankfurt am Main: Suhrkamp, 1985), 5:61: "Language does not arise piece-meal or atomistically, but it arises in all its parts immediately as a whole and thus organically [organisch]."

38. Although Schleicher basically advanced the same theory as in his *Sprachengeschichte*, he now felt perfectly comfortable describing language groups using biological classifications. See his *Die Sprachen Europas in systematischer Übersicht* (Bonn: König, 1850), 22–25, and 30.

39. Among his contemporaries, William Dwight Whitney dismissed Schleicher's conception of language as a law-governed, organic phenomenon. Whitney argued that actions produced by human will escaped the rule of law. See Whitney, "Schleicher and the Physical Theory of Language," 298–331 (note 20). This same kind of criticism has been voiced more recently. Eugen Seidel thinks Schleicher "erred" in regarding *Sprachwissenschaft* as a *Naturwissenschaft*, failing, as he supposedly did, to perceive the social character of language. See Eugen Seidel, "Die Persönlichkeit Schleichers," *Wissenschaftliche Beiträge der Friedrich-Schiller-Universität Jena* (1972): 8–17. Arsleff expresses a similar opinion (*From Locke to Saussure*, 294–95 [note 20]). Such judgments betray a poverty of historical understanding.

40. Schleicher published two articles in 1853 that employed a graphic illustration of a *Stammbaum*. One was in Czech, the other German. See, for instance, August Schleicher, "Die ersten Spaltungen des indogermanischen Urvolkes," *Allgemeine Zeitschrift für Wissenschaft und Literatur* (August 1853): 786–87.

41. Taub thinks that Friedrich Ritschl (1806–76), Schleicher's teacher at Bonn, may have suggested the tree-method of representation by his work in the establish-

ment of manuscript pedigrees. See Taub, "Evolutionary Ideas and 'Empirical' Methods," 185–86 (note 19).

42. See, for example, August Schleicher, *Die Deutsche Sprache* (Stuttgart: Cotta'scher Verlag, 1860), 58–59.

43. Ibid., 29 (note 42). From the beginning of his theorizing, Schleicher believed that common *Lautgesetze* (laws of oral expression) governed consonant and vowel changes of language families. In *Deutsche Sprache*, he began formulating macrolaws of language descent, such as the one mentioned earlier.

44. Ibid., 38 (note 42). William Dwight Whitney, commenting on such passages in *Deutsche Sprache* and comparable ones in *Bedeutung der Sprache*, vigorously dissented: "The rise of language had nothing to do with the growth of man out of an apish stock, but only with his rise out of savagery and barbarism. . . . Man was man before the development of speech began; he did not become man through and by means of it." See Whitney, "Schleicher and the Physical Theory of Language," 324–25 (note 20).

45. Whitney, "Schleicher and the Physical Theory of Language," 5 (note 20): "Speech is thus the expression of thought in sound, audible thought, just as, on the other hand, thought is inaudible speech."

46. See Richards, *Meaning of Evolution*, 42–55 (note 22).

47. Johann Gottfried Herder, *Abhandlung über den Ursprung der Sprache*, in *Sprachphilosophische Schriften*, ed. Erich Heintel (Hamburg: Felix Meiner, 1975), 3–90; quotations from pages 28 and 32.

48. *Thought*, according to Schleicher, has material elements—that is, representations (of phenomena) and concepts (when reflexive)—and formal structure—that is, the relationships among the elements. "Language thus has as its task to provide through sound an image of representations and concepts, and their relationships." *Meaning* (*Bedeutung*) then is the concept or representation as expressed in sound, while a word *root* is the sound complex that expresses meaning. The word itself is the meaning plus the grammatical relationships in sound. See Schleicher, *Die Deutsche Sprache*, 6 (note 42).

49. Humboldt, *Kawi-Sprache*, 1:lxxiv (note 35).

50. Ibid., 1:lxxiii (note 35).

51. Schleicher, *Zur vergleichenden Sprachengeschichte*, 11 (note 34).

52. Ibid., 16 (note 34). Schleicher quoted extensively from Hegel's *Introduction to the Philosophy of History*. This was part of the compilation of student notes published in 1840, after Hegel's death. Hegel maintained, for instance: "It is a fact, shown by literary remains, that the languages spoken by peoples in uncultured conditions have been well-formed in the highest degree, and that human understanding has developed through having this theoretical foundation. . . . It is further a fact that with the progressive civilizing of society and the state that the systematic activity of the understanding has eroded and language has become less well-formed and poorer." See Georg Wilhelm Friedrich Hegel, *Vorlesungen über die Philosophie der Geschichte*, vol. 12: *Werke*, 4th ed. (Frankfurt am Main: Suhrkamp, 1995), 85.

53. Hegel, *Vorlesungen*, 78 (note 52).

54. Ernst Haeckel, *Die Radiolarien*, 2 vols. (Berlin: Reimer, 1862).

55. I elaborate the history of Haeckel's development and the impact of romantic thought on his science in my forthcoming *The Tragic Sense of Life: Ernst Haeckel and the Struggle over Evolutionary Thought in Germany*.

56. See Ernst Haeckel, *Der Monismus als Band zwischen Religion und Wissenschaft* (Stuttgart: A. Kröner, 1905). Haeckel first explicitly endorsed Schleicher's conception of monism in his *Generelle Morphologie der Organismen*, 2 vols. (Berlin: Reimer, 1866), 1:105–108.

57. Ernst Haeckel, *Die Natürliche Schöpfungsgeschichte* (Berlin: Reimer, 1868), 550.

58. Ibid., 546 (note 57).

59. Ibid., 549 (note 57).

60. See Eugène Dubois, *Pithecanthropus erectus: Eine Menschenaehnliche Uebergangsform aus Java* (Batavia: Landesdruckerei, 1894).

61. Haeckel, *Natürliche Schöpfungsgeschichte*, 511 (note 57).

62. The debate over the monogenic or polygenic origin of man still rages, if in a slightly different key. See, for instance, Christopher Stringer and Robin McKie, *African Exodus: The Origins of Modern Humanity* (New York: Holt, 1996). See also my review of their book, "Neanderthals Need Not Apply," *New York Times Book Review* (August 17, 1997), 10.

63. Alec Panchen discusses such antecedents, for example, the "tree of Porphyry," in *Classification, Evolution, and the Nature of Biology* (Cambridge: Cambridge University Press, 1992), 10–40. He renders the obvious judgment that "the fashion for genealogical dendrograms, or phylogenetic trees, representing real taxa, started with Haeckel" (30).

64. See Richards, *Darwin and the Emergence of Evolutionary Theories*, 176–84 (note 1).

65. Wallace first advanced his arguments in a review of new editions of Charles Lyell's works. See Alfred Russel Wallace, "Review of *Principles of Geology* by Charles Lyell; *Elements of Geology* by Charles Lyell," *Quarterly Review* 126 (1869): 359–94.

66. Charles Darwin to Alfred Wallace (January 26, 1870), in *Alfred Russel Wallace: Letters and Reminiscences*, 2 vols., James Marchant, ed. (London: Cassell, 1916), 1:251.

67. Frederick Farrar, "Philology and Darwinism," *Nature* 1 (1870): 527–29.

68. Charles Darwin to William S. Dallas (June 9, 1868), in DAR 162, held in the Manuscript Room of Cambridge University Library.

69. Darwin's copy of Haeckel's *Natürliche Schöpfungsgeschichte* is held in the Manuscript Room of Cambridge University Library.

70. In the conclusion to the *Descent of Man*, Darwin referred to an article by Chauncy Wright, which in the last moments of manuscript preparation he had just read. Wright had attacked Wallace's argument that man's big brain had to be given a nonselectionist account. See Chauncy Wright, "Limits of Natural Selection," *The*

North American Review 111 (October 1870): 282–311. Darwin suggested that Wright also endorsed the idea that language operated to produce man's increased intellectual capacity through use inheritance (*Descent of Man*, 2:390–91 [note 5]). Wright's argument is a bit convoluted, but it is clear, he made no such argument as Darwin attributed to him. Quite the contrary. Wright (294–98) maintained that Wallace had simply misjudged the character of the native's capacities. Wright rather held that language and so-called higher faculties were merely collateral features of capacities directly useful to the native and so indirectly acquired through natural selection. "Why may it not be," he asked, "that all that he [the savage] can do with his brains beyond his needs is only incidental to the powers which are directly serviceable?" (295). He further suggested that the difference between the savage and the philosopher "depends on the external inheritances of civilization, rather than on the organic inheritances of the civilized man" (296). Darwin, in his enthusiasm for the Schleicher argument, found its ghost in any text that opposed Wallace's thesis.

Chapter 3

1. Ernst Robert Curtius, *Europäische Literatur und lateinisches Mittelalter* (Bern: A. Francke, 1948), 323.

2. Paul Berg and Maxine Singer, *Dealing with Genes: The Language of Heredity* (Mill Valley, Calif.: University Science Books, 1992), 241.

3. Bernd-Olaf Küppers, *Information and the Origin of Life* (Cambridge, Mass.: MIT Press, 1990), xix. Originally published in German: Bernd-Olaf Küppers, *Der Ursprung biologischer Information. Zur Naturphilosophie der Lebensentstehung* (München: Piper, 1986).

4. Umberto Eco, *Semiotics and the Philosophy of Language* (Bloomington: Indiana University Press, 1984), 87.

5. Ivo Braak, *Poetik in Stichworten* (Stuttgart: Borntraeger Verlag, 1990), 30.

6. Eco, *Semiotics and Philosophy of Language*, 101 (note 4). Aristotle, *Poetics* 1459 a6–8.

7. Erwin Schrödinger, *What Is Life?* (Cambridge: Cambridge University Press, 1944).

8. Lily E. Kay, "Who Wrote the Book of Life? Information and the Transformation of Molecular Biology, 1945–55," *Science in Context* 8 (1995): 609–34; Lily E. Kay, "A Book of Life? How the Genome Became an Information System and DNA a Language," *Perspectives in Biology and Medicine* 41 (1998): 504–28. See also Lily E. Kay, *Who Wrote the Book of Life? A History of the Genetic Code* (Stanford, Calif.: Stanford University Press, 2000), which appeared too late to be discussed here.

9. Norbert Wiener, *Cybernetics: Or Control and Communication in the Animal and the Machine* (New York: J. Wiley, 1948), 8; Norbert Wiener, *The Human Use of Human Beings: Cybernetics and Society* (Boston: Houghton Mifflin, 1950), 109.

10. Kay, "Who Wrote the Book of Life?" 624 (note 8).

11. Ibid., 625 (note 8).

12. H. B. Branson, "Information Theory and the Structure of Proteins," in *Essays on the Use of Information Theory in Biology*, ed. H. Quastler (Urbana: University of Illinois Press, 1953), 84–104.

13. George Gamow, "Possible Relation Between Deoxyribonucleic Acid and Protein Structures," *Nature* 173 (1954): 318.

14. Kay, "Who Wrote the Book of Life?"; Kay, "A Book of Life?" (note 8).

15. Vadim A. Ratner, *Molekulargenetische Steuerungssysteme* (Stuttgart: Fischer, 1977 [originally published in Russian, 1966]), 45. Eigen's arguments were correspondingly based on Chomsky's statement that language is organized on hierarchical and generative principles. Among other criteria formulated in modern linguistics, the linguists Jakobson and Raible seized on Chomskyan terms such as *hierarchical structure* and *recursivity*.

16. Ratner's terms are dated; they came out of work with prokaryotes and are not strictly applicable to eukaryotes.

17. Creativity, a most important characteristic of language, is linked to semantics; Chomsky has introduced the term *creativity* in linguistics; however, his generative properties of language limit this creativity to the syntactic level alone.

18. In Figure 3.1, a "morphe" is according to Ratner "the minimal connected sequence of letters in a text, having a certain meaning" (Ratner, *Molekulargenetische Steuerungssysteme*, 56 [note 15]). This corresponds to the notion of *morpheme* or *moneme*.

19. François Jacob et al., "Vivre et parler: Un débat entre François Jacob, Roman Jakobson, Claude Lévi-Strauss et Philipp L'Héritier," part 1, *Les lettres françaises* 1221 (February 14, 1968): 3–7; part 2, *Les lettres françaises* 1222 (February 21, 1968): 4–5. François Jacob and Jacques Monod, the founder of the operon model, also analyzed their biological work from a philosophical perspective. While Monod describes translation as a mechanical process and compares ribosomes to machines, Jacob uses mostly analogies and metaphors from computer science: "Today inheritance needs to be described in terms of information, messages and code." Jacques Monod, *Le hasard et la nécéssité* (Paris: Seuil, 1970); François Jacob, *La logique du vivant. Une histoire de l'hérédité* (Paris: Gallimard, 1970), 9.

20. Jakobson further mentioned binary oppositions, dialogue character, variability, stability, and legacy.

21. Roman Jakobson, "Relations entre la science du langage et les autres sciences," in his *Essais de Linguistique Générale II* (Paris: Les Éditions de Minuit, 1973), 9–76, 53.

22. Jacob et al., "Vivre et parler," part 1, 6 (note 19).

23. Ibid., part 1, 6 (note 19).

24. André Martinet, *Éléments de linguistique générale*, 2d ed. (Paris: Colin, 1961). Compare Jakobson, "Relations," 45 (note 21).

25. Martinet, in *Éléments*, speaks about "monemes." I will continue to use the

more widely used term *morpheme*. Morphemes correspond to the smallest linguistic signs defined by Saussure as units of expression as well as of content.

26. Jakobson, "Relations," 52 (note 21).

27. Ferdinand de Saussure, edited posthumously by Charles Bally and Albert Sechehaye, *Cours de linguistique générale*, 4th ed. (Paris: Payot, 1964–65 [1916]).

28. François Jacob, "Le modèle linguistique en biologie," *Critique* 322 (March 1974): 197–205, 201ff.

29. Jakobson, "Relations," 53 (note 21); Brian F. C. Clark, and Kjeld A. Marcker, "How Proteins Start," *Scientific American* 1 (1968): 36–42, quoted in Jakobson, "Relations," 68 (note 21).

30. Mitochondria provide a partial exception. In this case a sequence that usually serves as a stopping triplet codes an amino acid. However, there is no connection with the surrounding codons.

31. Jakobson, "Relations," 54 (note 21); Jacob, "Modèle linguistique en biologie," 198–99 (note 28).

32. Claude E. Shannon and Warren Weaver, *The Mathematical Theory of Communication*, 8th ed. (Urbana: University of Illinois, 1949), 104.

33. Manfred Eigen, "Sprache und Lernen auf molekularer Ebene," in *Der Mensch und seine Sprache*, ed. Carl Friedrich von Siemens Stiftung (Berlin: Propyläen Verlag, 1979), 181–218; Manfred Eigen and Ruthild Winkler, *Das Spiel. Naturgesetze steuern den Zufall* (München: Piper, 1975); English edition: *Laws of the Game: How the Principles of Nature Govern Chance* (New York: Alfred A. Knopf, 1981).

34. Eigen, "Sprache und Lernen," 198 (note 33).

35. Ibid., 192–93 (note 33).

36. Eigen and Winkler, *Laws of the Game*, 272 (note 33).

37. Eigen, "Sprache und Lernen," 304 (note 33).

38. Ibid., 304 (note 33).

39. Eigen and Winkler, *Laws of the Game*, 280 (note 33).

40. Eigen, "Sprache und Lernen," 181 (note 33).

41. Eigen and Winkler, *Laws of the Game*, 271 (note 33). In this quotation I have replaced the original translation "objective language" with "object-language," which corresponds better to the term "Objektsprache" used by Eigen and Winkler in *Das Spiel*, 304 (note 33).

42. Eigen, "Sprache und Lernen," 181 (note 33). When Eigen's pupil Küppers asserts that the notion of "molecular language" is more than a simple metaphor, this statement then refers only to an abstract notion of *language* and not to the specific phenomenon of *human language* with all its characteristics. Küppers, *Information and the Origin of Life*, 21 (note 3).

43. Eigen and Winkler, *Laws of the Game*, 274 (note 33); Eigen, "Sprache und Lernen," 185 (note 33).

44. Eigen, "Sprache und Lernen," 196 (note 33).

45. Ibid., 184 (note 33).

46. Ibid., 181 (note 33).

47. Jakobson conceived language as a means of transmission and also as the substrate of learning, using the term "legacy" to describe it: "language as heredity, as a testament, as instruction, coming from the past and going to the future" (Jacob et al., "Vivre et parler," part 1, 6 [note 19]). The "genetic" and human language have the same function: both are memories of humanity.

48. Eigen and Winkler, *Laws of the Game*, 260 (note 33).

49. Wolfgang Raible, "Sprachliche Texte—genetische Texte: Sprachwissenschaft und molekulare Genetik," *Sitzungsberichte der Heidelberger Akademie der Wissenschaften/Philosophisch-historische Klasse: Bericht* (1993): 1–66.

50. Ibid., 21 (note 49).

51. Ibid., 44 (note 49).

52. Ibid., 39 (note 49). Raible uses the term "form class" synonymously with "word class," as in many modern works.

53. Ibid., 40 (note 49).

54. Ibid., 38 and 43 (note 49).

55. Ibid., 39 (note 49).

56. I will deal with the notion of *communication* and with the notion of *sense* in more detail below, however, it should be introduced here also in connection with the question of whether homeotic genes have a metacommunicative function.

57. Raible, "Sprachliche Texte—genetische Texte," 45 (note 49).

58. Ibid., 59 (note 49).

59. Ibid., 51 (note 49).

60. Ibid., 46 (note 49).

61. Ibid. (note 49).

62. Noam Chomsky, *Syntactic structures* (The Hague: Mouton, 1957).

63. Raible, "Sprachliche Texte—genetische Texte," 15 (note 49).

64. Contrary to Raible, Martinet proceeded not from the smaller unit to the larger, but inversely ordered the monemes to the first articulation, the phonemes to the second articulation (Martinet, *Linguistique générale*, 13–14 [note 24]).

65. Eco, *Semiotics and Philosophy of Language*, 183 (note 4).

66. Ibid., 183 (note 4).

67. Jacob, "Modèle linguistique en biologie," 198 (note 28).

68. Eugenio Coseriu, "Bedeutung und Bezeichnung im Lichte der strukturellen Semantik," in *Sprachwissenschaft und Übersetzen*, ed. P. Hartmann and H. Vernay (Munich: Hueber, 1970), 104–21, 105.

69. Raible, "Sprachliche Texte—genetische Texte," 54 (note 49).

70. Raible, "Sprachliche Texte—genetische Texte," 37 (note 49).

71. Ibid., 26 (note 49).

72. Ibid., 45 (note 49).

73. Ibid., 20 (note 49).

74. For example, the sentence "I am cold" follows certain specific linguistic rules and signifies a "freezing person"; only the context can tell what is meant by the statement: for example, to close the window.

75. In relation to cell differentiation, Küppers also remarks about the notion of *context*: "The recipient of the information is in the medium in the cell that surrounds the DNA. Only in the context of the specific physico-chemical milieu of the cell can the information encoded in the DNA develop to realise the function to which it corresponds." This definition of context erases the differentiation between receiver and context, a situation impossible in human language. Bernd-Olaf Küppers, "The Context-Dependence of Biological Information" in *Information: New Questions to a Multi- Disciplinary Concept*, ed. Klaus Kornwachs and Konstantin Jacoby (Berlin: Akademie-Verlag, 1996), 137–45, 142–43.

76. Karl Bühler, *Sprachtheorie*, 2d ed. (Stuttgart: G. Fischer, 1965), 24–33. Roman Jakobson, "Linguistics and Poetics," in *Style in Language*, ed. Th. A. Sebeok (Cambridge, Mass.: MIT Press, 1960), 350–77.

77. The "vectorial character of language exchange from the chromosome to the organism" (Eigen and Winkler, *Laws of the Game*, 306 [note 33]) is expressed in the "central dogma" of biology: *DNA makes RNA makes protein.*

78. Küppers, "Information and the Origin of Life," 77 (note 3).

79. An advantage of this comparison is that it encompasses, on the one hand, the observation that it is possible to express the genetic information of biological macromolecules in the form of a binary code, which is not possible for the human brain and seems to explain the limits of computer technology (Küppers, "Information and the Origin of Life," 21 [note 3]); and on the other, the fact that computer science can "make use of the mechanisms of evolution" (John H. Holland, "Genetische Algorithmen," *Spektrum der Wissenschaft* [September 1992]: 44–51, 44). Computer programs known as genetic algorithms are based on the fundamental processes of evolution. For discussion of the controversy over the information metaphor and its attendant terms such as *program*, see T. Fogle, "Information Metaphors and the Human Genome Project," *Perspectives in Biology and Medicine* 38 (1995): 535–47.

80. Charles F. Hockett, "The Problem of Universals in Language," in *Universals of Language*, 2d ed., ed. J. Greenberg (Cambridge, Mass.: MIT Press, 1965), 1–29, 15.

81. Ibid., 15 (note 80).

82. Raible, "Sprachliche Texte—genetische Texte," 50 (note 49).

83. Another approach toward a definition of language is the study of genericessential universals, defined by Oesterreicher as "constituents of the notion of language itself that is aspects which are rationally necessary to define language." Oesterreicher's universal of historicity (*Historizität*) regards language as something that has emerged over time and undergoes constant changes. Oesterreicher understood the historicity of language as the result of a "balance" between two universal principles: *Alterität* and creativity. The first takes into account the aim for linguistic concurrence, necessary for a mutual understanding and contributing to the stability of language. The second corresponds to the desire of the speaking person to constantly create new forms of expressions and communication, explaining historical changes in languages. Wulf Oesterreicher, *Sprachtheorie und Theorie der Sprachwissenschaft*

(Heidelberg: Winter, 1979), 226 and 247. Similar to Oesterreicher, Eigen argued that language is characterized by stability and generativity (*Generativität*), whereas in evolution there is also a balance between two principles: functionality and creativity. Eigen, "Sprache und Lernen," 206 (note 33). For Raible, however, linguistic creativity, which he opposes to the biological phenomenon of replication, is the most important difference between language and the genetic realm. Raible, "Sprachliche Texte —genetische Texte," 54 (note 49).

84. Jacob et al., "Vivre et parler," part 1, 6 (note 19).

85. Jacob, "Modèle linguistique en biologie," 200 (note 28).

86. Raible, "Sprachliche Texte—genetische Texte," 43 (note 49).

87. Ibid., 17 (note 49).

88. Jacob et al., "Vivre et parler," part 1, 6–7 (note 19).

89. Küppers, "Information and the Origin of Life," 171 (note 3).

90. Jürgen von Stenglin, *Denken der Wirklichkeit. Eine sprachlich und kognitiv fundierte Theorie der Erkenntnis* (Würzburg: Königshausen & Neumann, 1990), 87.

91. Ibid., 87 (note 90).

92. Oesterreicher, *Sprachtheorie und Theorie der Sprachwissenschaft*, 251 (note 83).

93. Stenglin, *Denken der Wirklichkeit*, 148 (note 90).

94. Carl Friedrich von Weizsäcker, *Die Einheit der Natur*, 4th ed. (Munich: Hanser, 1972), 51–52.

95. Adam Hedgecoe does not see a connection between the notions of *information* and *language*: "This article argues that the linguistic metaphor—e.g. 'the language of the genes'—should be seen as being separate from the information metaphor with which it is usually associated." He justifies this by saying that the "metaphor of information" does not reach as far as the "metaphor of language." So a deeper analogy between the genetic apparatus and language could be justified on the basis of a certain definition of language: "It is possible to view language metaphor as transformed, . . . but this would involve a different interpretation of language." However, "information" is a dead metaphor for Hedgecoe: it is impossible to speak of DNA as a "code" in the sense of information theory or cybernetics, because in these the notion of information is "explicitly independent of meaning and function." In fact, geneticists would not use the notion of information based on Shannon and Weaver's definition: while "'genetic information' might have been a metaphoric term in the 1950s, its current use in both scientific and popular discourse about genetics is so widespread that it can be seen as a 'dead metaphor,' i.e. independent of the original meaning, having taken on a separate existence of its own." I disagree with Hedgecoe, both because the notion of information borrows from the notion of language, and because one cannot speak of "meaning" in the genetic realm. Although the term *information* is used in the popular realm, one cannot deny that also in the genetic realm there is safeguarding and transmission of information, and—as Hedgecoe himself in fact recognizes—there is no general definition for the notions either of information or of language. Adam M. Hedgecoe, "Transforming Genes: Metaphors

of Information and Language in Modern Genetics," *Science as Culture* 8 (1999): 209–29, 209, 224, 218, and 221–22.

96. Kay, "A Book of Life?" 510–11 (note 8).

97. R. N. Mantegna, S. V. Buldyrev, A. L. Goldberger, S. Havlin, C. K. Peng, and H. E. Stanley, "Linguistic Features of Noncoding DNA Sequences," *Physical Review Letters* 73 (1994): 3169–72.

98. George K. Zipf, *Human Behaviour and the Principle of Least Effort* (Cambridge, Mass.: Addison-Wesley Press, 1949 [republished in facsimile, 1972]), 19–20.

99. Faye Flam, "Hints of a Language in Junk DNA," *Science* 266 (1994): 1320.

100. Küppers, "Information and the Origin of Life," 93 (note 3).

101. Quoted in Flam, "Hints of a Language," 1320 (note 99).

102. Hedgecoe, "Transforming Genes," 224 (note 95).

103. S. Bonhoeffer, A. V. M. Herz, M. C. Boerlijst, S. Nee, M. A. Nowak, and R. M. May, "No Signs of Hidden Language in Noncoding DNA," *Physical Review Letters* 76 (1996): 1977.

104. R. F. Voss, "Comment on Linguistic Features of Noncoding DNA Sequences," *Physical Review Letters* 76 (1996): 1978.

105. N. E. Israeloff, M. Kagalenko, and K. Chan, "Can Zipf Distinguish Language from Noise in Noncoding DNA?" *Physical Review Letters* 76 (1996): 1976.

106. G. S. Attard, A. C. Hurworth, and J. P. Jack, "Language-Like Features in DNA: Transposable Element Footprints in the Genome," *Europhysics Letters* 36 (1996): 391–96, 396.

107. A. A. Tsonis, J. B. Elsner, and P. A. Tsonis, "Is DNA a Language?" *Journal of Theoretical Biology* 184 (1997): 25–29, 27.

108. Ibid., 29 (note 107).

109. Patricia Bralley, "An Introduction to Molecular Linguistics," *BioScience* 46 (1996): 146–53.

110. R. Boyd, "Metaphor and Theory Change: What Is 'Metaphor' a Metaphor For?" in *Metaphor and Thought*, 2d ed., ed. S. A. Ortony (Cambridge: Cambridge University Press, 1993), 481–532, 486. See also Hedgecoe, "Transforming Genes," 212 (note 95).

111. Richard Rorty, *Contingency, Irony, and Solidarity* (Cambridge: Cambridge University Press, 1989), 8–9.

112. Hedgecoe, "Transforming Genes," 213 (note 95).

113. See Evelyn Fox Keller's chapter in this volume.

114. Jacob, "Modèle linguistique en biologie," 203–204 (note 28).

115. Ibid., 204 (note 28). See also Hans-Jörg Rheinberger, "Konjunkturen, Transfer-RNA, Messenger-RNA, genetischer Code," in *Objekte, Differenzen und Konjunkturen*, ed. M. Hagner, H.-J. Rheinberger, and B. Wahrig-Schmidt (Berlin: Akademie-Verlag, 1994), 201–31.

116. Hedgecoe, "Transforming Genes," 224 (note 95).

117. See Evelyn Fox Keller's chapter in this volume.

Chapter 4

1. A. Weismann, *The Evolutionary Theory*, 2 vols., trans. J. A. and M. R. Thomson (London: Edward Arnold, 1904), 2:63.

2. Ibid., 2:107 (note 1), quoted in John Maynard Smith, "Weismann and Modern Biology," in *Oxford Surveys in Evolutionary Biology*, ed. P. H. Harvey and L. Partridge (Oxford: Oxford University Press, 1989), 6:1–12, 7.

3. Walter Benjamin, "The Task of the Translator," (1923), reprinted in *Illuminations* (New York: Harcourt, Brace and World, 1968), 69–82, 72.

4. Weismann, *Evolutionary Theory*, 108 (note 1).

5. James R. Griesemer and William C. Wimsatt, "Picturing Weismannism: A Case Study of Conceptual Evolution," in *What the Philosophy of Biology Is*, ed. Michael Ruse (Dordrecht: Kluwer Academic Publishers, 1989).

6. Ibid. (note 5).

7. Francis Crick, "On Protein Synthesis," *Symposium of the Society of Experimental Biology* 12 (1957): 138–63, 153. Crick's emphasis.

8. Francis Crick, "Central Dogma of Molecular Biology," *Nature* 227 (1970): 461–563, 562.

9. J. Maynard Smith, *The Theory of Evolution*, 2d ed. (Middlesex: Penguin, 1965).

10. Maynard-Smith, "Weismann and Modern Biology," 7 (note 2).

11. Crick, "Protein Synthesis," 161 (note 7). Twelve years later, Crick recollected, "All we had to work on were certain fragmentary experimental results, themselves often rather uncertain and confused, and a boundless optimism that the basic concepts were rather simple . . . " (Crick, "Central Dogma," 561 [note 8]).

12. Robert Sinsheimer, "The Prospect of Designed Genetic Change," *Engineering and Science* 32 (1969): 8–13, 12; reprinted in Ruth Chadwick, ed., *Ethics, Reproduction, and Genetic Control* (London: Croom Helm, 1987), 136–46.

13. Francis Crick, "Eugenics and Genetics," in *Man and His Future*, ed. G. Wolstenholme (Boston: Little, Brown and Co., 1963), 275.

14. Crick, "Eugenics and Genetics," 284 (note 13).

15. Ibid., 295 (note 13).

16. Robert Cook-Deegan, *The Gene Wars: Science, Politics, and the Human Genome* (New York: W. W. Norton and Co., 1994), 161.

17. This is perhaps especially clear in the recent shifts to corporate funding for a number of genomics projects (for example, Craig Venter's efforts to sequence the human genome).

18. P. E. Griffiths and R. D. Gray, "Developmental Systems and Evolutionary Explanation," *Philosophy of Science* 91 (1994): 277–304, 284.

Chapter 5

This chapter is adapted from *Astrology and the Seventeenth Century Mind: William Lilly and the Language of the Stars* (Manchester: Manchester University Press, 1995).

1. P. L. Berger and T. Luckmann, *The Social Construction of Reality* (New York: Doubleday, 1967), esp. 113–15.

2. E. Garin, *Astrology in the Renaissance: The Zodiac of Life* (London: Routledge & Kegan Paul, 1983), 25. (Garin says that he disagrees with Boll on this point.)

3. C. H. Josten, ed., *Elias Ashmole*, 5 vols. (Oxford: Clarendon Press, 1966), 1:21; L. Thorndike, "The True Place of Astrology in the History of Science," *Isis* 46 (1955): 273–78, 273.

4. W. Lilly, *Annus Tenebrosus* (London, 1652), preface; cited to "Albertus Mag. tom. 5, pag. 659": "Universi ordinationem nulla scientia humana perfecte attingit, sicut scientia judiciorum Astrorum."

5. See Geneva, *Astrology and the Seventeenth Century Mind*, for the background to this claim.

6. J. Dryden, *An Evening's Love, or The Mock-Astrologer* (London, 1671), 15.

7. J. Milton, *Paradise Lost*, book 6, lines 312–15.

8. J. Aubrey, *Brief Lives*, 2 vols., ed. Andrew Clark (Oxford: Clarendon Press, 1898 [1669–96]), 1:372.

9. W. Laud, *A Sermon preached on Munday, the Seaventeenth of March, at Westminster: At the opening of the Parliament* (London, 1628), 22. (Note Laud's emphasis on the ascendant.) John Gadbury in the preface to his 1684 *Cardines Coeli*, which contained ten genitures including Laud's, used this sermon to argue Laud's familiarity with and belief in astrology. The *Centiloquium's Aphorisms* were mainstays of seventeenth-century astrology, although Ptolemy's authorship was ultimately discredited.

10. *The Journals of the House of Commons*, vol. 3, (Reprint by Order of The House of Commons, 1803), entry for 13 September 1660.

11. J. Howell, *Dodona's Grove*, 2d ed. (London, 1644), 141–42.

12. Anon., *The English Fortune-Teller* (1642), 3.

13. Ibid., 5 (note 12).

14. J. Drucker, *The Alphabetic Labyrinth* (London: Thames and Hudson, 1995), 123–25.

15. K. Thomas, *Religion and the Decline of Magic* (New York: Scribner, 1971), 375.

16. K. Thomas, "University of Oxford Examination Schools," 31 October 1986.

17. Plotinus, *Ennead* 2:3. "On Whether the Stars are Causes," trans. A. H. Armstrong (Cambridge, Mass.: Harvard University Press, 1966), 69.

18. G. Atwell, *An Apology, Or, Defence of the Divine Art of Natural Astrologie* (London, 1660), 8.

19. Ms. Ashmole 186, f. 141v. (Bodleian Library, Oxford).

20. W. Lilly, *An Astrologicall Prediction of the Occurrances in England, Part of the Years 1648, 1649, 1650* (London, 1648), 18.

21. W. Lilly, *Anglicus 1649* (London, 1648), 'To the Reader,' sig. A4.

22. W. Lilly, *The Starry Messenger* (London, 1645), 41.

23. For a much more detailed discussion of this multilayered astrological syntax, see Geneva, *Astrology and the Seventeenth Century Mind*.

24. J. Melton, *Astrologaster or the Figure-Caster*, ed. H. G. Dick (Los Angeles: W. A. Clark Memorial Library, 1975 [1620]), 65.

25. Ibid., 20 (note 24).

26. Ibid., 15 (note 24).

27. B. Willey, *The Seventeenth Century Background: Studies in the Thought of the Age in Relation to Poetry and Religion* (London: Chatto & Windus, 1934), 119.

28. V. Salmon, *The Works of Francis Lodwick* (London: Longman, 1972), 76.

29. For Bacon's influence on the universal language movement, see M. M. Slaughter, *Universal Languages and Scientific Taxonomy in the Seventeenth Century* (Cambridge: Cambridge University Press, 1982), esp. 90–97; and Salmon, *Francis Lodwick*, esp. 13–15 (note 28).

30. J. Wilkins, *Mercury, or the Secret and Swift Messenger* (London, 1641), 108.

31. H. Asbach-Schnitker, ed., *John Wilkins, Mercury, or the Secret and Swift Messenger*, 3d ed. (facsimile reprinted Amsterdam: J. Benjamin's Pub. Co., 1984 [1708]), x.

32. Salmon, *Francis Lodwick*, 22 (note 28).

33. T. Sprat, *History of the Royal-Society of London, For the Improving of Natural Knowledge*, ed. J. I. Cope and H. W. Jones (St. Louis, Mo.: Washington University Press, 1958 [1667]), f. B2; J. A. Winn, *John Dryden and His World* (New Haven, Conn.: Yale University Press, 1987), 131.

34. Salmon, *Francis Lodwick*, 30 (note 28).

35. S. Ward, *Vindiciae Academiarum* (London, 1654), 22. Most probably he was referring to Wilkins.

36. W. P. D. Wightman, *Science in a Renaissance Society* (London: Hutchinson, 1972), 68.

37. Willey, *Seventeenth Century Background*, 133 (note 27).

38. J. Webster, *Academiarum Examen* (London, 1654), 8.

39. Ibid., 25 (note 38).

40. Ibid., 28 (note 38).

41. E. Ashmole, *Fasciculus Chemicus* (London, 1650), "Prolegomena," 29–30; Josten, *Elias Ashmole*, 2:68–69 (note 3).

42. Webster, *Academiarum Examen*, 119 (note 38).

43. Salmon, *Francis Lodwick*, 53 (note 28).

44. Slaughter, *Universal Languages and Scientific Taxonomy*, viii (note 29).

45. Ibid., vii (note 29).

46. J. Wilkins, *Vindiciae Academiarum* (London, 1654), 5.

47. Ward, *Vindiciae Academiarum*, 17–18 (note 35).

48. Ibid., 13 [wrongly paginated as 5] (note 35).

49. Ibid., 18 (note 35).

50. Sprat, *History of the Royal-Society*, 5 (note 33).

51. Ibid., 97 (note 33).

52. Ibid., 42 (note 33).

53. Webster, *Academiarum Examen*, 51 (note 38).

54. Ward, *Vindiciae Academiarum*, 30 (note 35). Ward does, however, admit to having respect for one practitioner of the art—Mr. Ashmole.

55. Ibid., 46 (note 35).

56. Ibid., 19 (note 35).

57. See Slaughter, *Universal Languages and Scientific Taxonomy*, 110–11 (note 29); Salmon, *Francis Lodwick*, 16 (note 28). Wit-Spell had been written by a Rev. Johns[t]on at the request of Bishop William Bedell of Kilmore, one of Hartlib's friends.

58. G. W. Leibniz, *Opera philosophica*, 2 vols., ed. J. E. Erdman (Berlin: G. Eichleri, 1839–40); as summarized in F. Yates, *The Art of Memory* (Chicago: Chicago University Press, 1966), 382–85. He, however, went further down the road toward the ultimate universal scientific language—mathematics—in a fragment *Lingua generalis*, in which he argues that "complex notions may be expressed as easily as large numbers, if only an order can be established among concepts so that everything may be represented clearly to human judgment, simple concepts being denoted by primary numbers" (Salmon, *Francis Lodwick*, 41 [note 28]).

59. F. Bacon, *Advancement of Learning* in *The Works of Francis Bacon*, 14 vols., ed. J. Spedding et al. (London: Longman and Co., 1857–74), 3:396–97.

60. Galileo Galilei, *Assayer*, trans. Stillman Drake, in A. M. Smith, "Knowing Things Inside Out—The Scientific Revolution from a Medieval Perspective," *American Historical Review* 95 (1990): 726–44, 737.

61. T. Hobbes, *Leviathan*, ed. A. R. Waller (Cambridge: Cambridge University Press, 1935), 1:4, 17.

62. Sprat, *History of the Royal-Society*, 40 (note 33).

63. Ibid., 111–15 (note 33).

64. Ibid., 113; emphasis mine (note 33). The earlier seventeenth century had made Cato's "Rem tene, verba sequentur" a standard for its ideal of balanced brevity (see Cope and Jones in Sprat, *History of the Royal-Society*, xxix [note 33]).

65. A. C. Crombie, *Augustine to Galileo* (London: William Heinemann, 1957), 74.

66. Slaughter, *Universal Languages and Scientific Taxonomy*, 6 (note 29).

67. Sprat, *History of the Royal-Society*, 325 (note 33).

68. See Mordechai Feingold, "The Occult Tradition in the English Universities of the Renaissance: A Reassessment," in *Occult and Scientific Mentalities in the Renaissance*, ed. Brian Vickers (Cambridge: Cambridge University Press, 1984), 73–94, 82–83.

69. W. Lilly, *Englands Propheticall Merline* (London, 1644), 94.

70. R. P. Warren, "The Mission," in his *Now and Then, Poems 1976–78* (New York: Random House, 1978).

Chapter 6

All translations from French manuscripts and texts are mine, and I am responsible for any inaccuracy.

1. Antoine Ferchaut de Réaumur, "Règles pour construire des thermomètres dont les degrés soient comparables," *Mémoires de l'Académie Royale des Sciences* (Paris, 1730): 452–507, 453.

2. Arthur Birembaut, "La contribution de Réaumur à l'histoire de la thermométrie," in *La vie et l'œuvre de Réaumur (1683–1757)*, ed. P. P. Grassé (Paris: Presses Universitaires de France, 1962), 43–69; Maurice Daumas, *Les instruments scientifiques au XVIIème et XVIIIème siècles* (Paris: Presses Universitaires de France, 1953).

3. Jean Dominique Maraldi, "Observations du baromètre, et du thermomètres faites en différentes villes pendant l'année 1705," *Mémoires de l'Académie royale des Sciences* (Paris, 1706): 12–13; Bernard Le Bovier de Fontenelle, "Comparaison d'observations faites en différents lieux sur le Baromètre, sur les vents, et sur la quantité des pluyes," *Histoire de l'Académie Royale des Sciences* (Paris, 1706): 20–22.

4. William Derham, "Tables of the Barometrical Altitudes at Zurich in Switzerland in the year 1708 observed by Dr. J. J. Scheuchzer, F.R.S., and at Upminster in England, observed at the same time by Mr. W. Derham, F.R.S., as also the rain at Pisa in Italy in 1707 and 1708, observed there by Dr. Michael Angelo Tilli, and at Zurich in 1708, and at Upminster in all that time: with remarks on the same tables, as also on the Winds, Heat and Cold, and divers other Matters occurring in those different parts of Europe," *Philosophical Transactions of the Royal Society* 26 (1709): 342–66; "An abstract of the Meteorological Diaries, communicated to the Royal society, with remarks upon them," *Philosophical Transactions of the Royal Society* 38 (1734): 101–109, 334–44.

5. James Jurin, "Invitatio ad Observationes Meteorologicas communi consilio instituendas," *Philosophical Transactions of the Royal Society* 32 (1723): 422–27.

6. Guillaume Amontons, "Discours sur quelques propriétés de l'air, et le moyen d'en connaître, et le moyen d'en connaître la température dans tous les climats de la terre," *Mémoires de l'Académie Royale des Sciences* (Paris, 1702): 161–80; Bernard le Bovier de Fontenelle, "Sur une nouvelle propriété de l'air et une nouvelle construction du thermomètre," *Mémoires de l'Académie Royale des Sciences* (Paris, 1702): 1–8.

7. Steven Shapin, *A Social History of Truth* (Chicago: University of Chicago Press, 1994).

8. Josef Konvitz, *Cartography in France 1660–1848: Science, Engineering and Statecraft*, (Chicago: University of Chicago Press, 1987).

9. Christin, "Description de la méthode d'un thermomètre universel," *Mémoires de Trévoux* (Paris, 1742): 992–1002; Christin, "Observations sur la Méthode d'un Thermomètre Universel, lues à l'Académie des Beaux-Arts de Lyon," *Mémoires de Trévoux* (Paris, 1743): 197–221.

10. Antoine Ferchaut de Réaumur, "Règles," 452–53 (note 1).

11. Two letter fragments from Granger can be found in the Réaumur files at the Archives de l'Académie des Sciences.

12. Bernard le Bovier de Fontenelle, "Sur une irrégularité de quelques baromètres," *Histoire de l'Académie Royale des Sciences* (Paris, 1705): 16–21; Bernard le Bovier de Fontenelle, *Histoire de l'Académie Royale des Sciences* (Paris, 1706): 1–3; Guillaume Amontons, "De la hauteur du mercure dans les baromètres," *Mémoires de l'Académie Royale des Sciences* (Paris, 1705): 229–35; Guillaume Amontons,

"Suite des remarques sur la hauteur du mercure dans les baromètres," *Mémoires de l'Académie Royale des Sciences* (Paris, 1705): 267–74.

13. Christin, "Sur la méthode," 999 (note 9).

14. Ibid., 999 and 1002 (note 9).

15. Pierre-Mathias De Gourné, *Géographie Méthodique* (Paris, 1741), lxxvi.

16. Norton Wise, ed., *The Values of Precision* (Princeton, N.J.: Princeton University Press, 1995); Joseph O'Connell, "Metrology: The Creation of Universality by the Circulation of Particulars," *Social Studies of Science* 23 (1993): 129–73.

17. Antoine Ferchaut de Réaumur, "Règles," 501 (note 1).

18. John Greenberg, "Degrees of Longitude and the Earth's Shape: The Diffusion of a Scientific Idea in Paris in the 1730s," *Annals of Science* 41 (1984): 151–58; John Greenberg, "Geodesy in Paris in the 1730s and the Paduan Connection," *Historical Studies in the Physical Sciences* 13 (1983): 238–60; Rob Iliffe, "Aplatisseur du monde et de Cassini: Maupertuis, Precision Measurement, and the Shape of the Earth in the 1730s," *History of Science* 31 (1993): 335–75; Mary Terrall, "Representing the Earth's Shape: The Polemics Surrounding Maupertuis' Expedition to Lapland," *Isis* 83 (1992): 218–37; Antonio Lafuente and Antonio Delgado, *La geometrización de la tierra: Observaciones y resultados de la expedición geodésica hispano-francesa al Virreinato del Peru (1735–1744)* (Madrid: Instituto Arnau de Vilanova, 1984).

19. Antonio de Ulloa, *Voyage historique de l'Amérique méridionale fait par ordre du Roi d'Espagne* (Paris, 1752), 181.

20. Pierre Bouguer, *Journal de Voyage: Description de la Cordillière du Pérou*, undated manuscript, Archives de l'Observatoire, Paris, file C27-8.

21. Jacques Auguste de Thou, *Histoire Universelle* (Basel, 1742), 11:46–47.

22. Simon Schama, *Landscape and Memory* (New York: Alfred A. Knopf, 1995), part III.

23. Robert Boyle, "New Experiments Physico-Mechanical touching the Spring of the Air," in *Works* (London, 1727), 227.

24. Norman Bryson, *Word and Image: French Painting of the Ancien Regime* (Cambridge: Cambridge University Press, 1981).

25. As in the case of mountain travels, the use of instruments sharply separated spectacular profit-raising ascents from properly "scientific" ones allegedly useful to the advancement of science. See, for instance, the works of Robertson, who was always trying to negotiate his own stance on one side or the other of this razor's edge: Etienne-Gaspard Robertson, *Mémoires récréatifs, scientifiques et anecdotiques d'un physicien-aéronaute*, 2 vols. (Paris, 1831–33), 2:15, 65.

26. Barthelemi Faujas de Saint-Fond, *Description de la machine aérostatique de M. de Montgolfier et de celles auxquelles cette découverte à donné lieu*, 2 vols. (Paris, 1783), 1:48.

27. Faujas de Saint-Fond, *Description*, 53 (note 26).

28. On balloon ascents in late-eighteenth-century France, see Richard Gillespie, "Ballooning in France and Britain, 1783–1786. Aerostation and Adventurism," *Isis* 75 (1984): 249–68; Maurice Crosland, *Gay-Lussac, Scientist and Bourgeois* (Cam-

bridge: Cambridge University Press, 1978), 28–30. For an analysis of Robertson's practice and ascents, see Thomas L. Hankins and Robert J. Silverman, *Instruments and the Imagination* (Princeton, N.J.: Princeton University Press, 1995), chap. 3. On the importance of meteorology and balloon ascents to peripheral *sociétés savantes*, see James Mc Clellan III, *Colonialism and Science: Saint Domingue in the Old Regime* (Baltimore, Md.: John Hopkins University Press, 1992), chap. 10.

29. Gay-Lussac, "Relation d'un voyage fait par M. Gay-Lussac le 29 Fructidor an XII, et lu à l'Institut National, le 9 Vendémiaire an XIII," *Journal de Physique* 59 (1804): 454–62, 456–58; see also Jean-Baptiste Biot, "Relation d'un Voyage Aero-statique fait par M. Gay-Lussac et Biot, lue à la Classe des sciences mathématiques et physique de l'Institut National, le 9 Fructidor an XII," *Journal de Physique* 59 (1804): 314–20, 315.

30. Steven Shapin and Simon Schaffer, *Leviathan and the Air-Pump: Hobbes, Boyle, and the Experimental Life* (Princeton, N.J.: Princeton University Press, 1985).

31. One just has to look among Delille's many manuscripts conserved in the archives of the Observatoire, such as his "Instructions sur la manière de faire les observations météorologiques qui se puissent comparer avec celles qui se font depuis quelques années en plusieurs endroits de la Russie" (1732), in file A7-3. Delille, "Observations du thermomètre et voyages dans les contrées froides," Archives de l'Observatoire, Paris, file A7-1.

32. Anthony Giddens, *The Consequences of Modernity* (Stanford, Calif.: Stanford University Press, 1990).

Chapter 7

1. In this chapter I will not consider the physical side of computer hardware.

2. Edsger Dijkstra, a proponent of a narrow mathematical point of view, who had some influence on the self-understanding of computer science, considers programming nothing but "Very Large Scale Application of Logic." This opinion is all the more surprising, as Dijkstra insists in the same article that modeling with the computer is something radically new, which requires command of a wide range of semantic levels: "The programmer . . . has to be able to think in terms of conceptual hierarchies that are much deeper than a single mind ever needed to face before. Compared to that number of semantic levels, the average mathematical theory is almost flat. By evoking the need for deep conceptual hierarchies, the automatic computer confronts us with a radically new intellectual challenge that has no precedent in our history." Dijkstra's strangely contradictory attitude seems to arise from his desire to keep the discipline of computer science "pure" and to build a "firewall" against the social sciences. E. W. Dijkstra, "On the Cruelty of Really Teaching Computing Science," in "A Debate on Teaching Computer Science," ed. P. J. Denning, *Communications of the ACM* 32, no. 12 (1989): 1398–1404, 1402 and 1400.

3. Yet this formal analogy is constitutive for most language analogies in the natural sciences, which refer essentially only to the rule-based productive mechanism of

grammars elaborated in mathematically oriented linguistics. See Suhr's chapter in this volume.

4. Here I see the main difference from language analogies in the natural sciences, characterizing computer science as a linguistic technique or as "technical semiotics." Even formally oriented linguists emphasize the nonnatural character of language, which prevents its analysis with information-theoretical terms. Roman Jakobson argued in 1961 for the social status of language as a means of communication: "Attempts to construct a model of language without any relation either to the speaker or the hearer and thus to hypostasize a code detached from actual communication threaten to make a scholastic fiction from language." Speech communities and the use of language are required in a theory of communication, while in nature there is information, at most. "The interlocutors belonging to one given speech community may be defined as actual users of one and the same language code encompassing the same legisigns. A common code is their communication tool, which actually underlies and makes possible the exchange of messages. Here is the essential difference between linguistics and the physical sciences." Stretching this position a little, I would like to show in this chapter that formal languages can constitute "speech communities," even if they are based on a rather restricted code. R. Jakobson, "Linguistics and Communication Theory," in *Proceedings of the Twelfth Symposium in Applied Mathematics: Structure of Language and its Mathemathical Aspects*, ed. R. Jakobson (Providence, R.I.: American Mathematical Society, 1961), 245–52, 250 and 247.

5. Wilhelm von Humboldt, *On Language: The Diversity of Human Language-Structure and its Influence on the Mental Development of Mankind*, trans. Peter Heath (Cambridge: Cambridge University Press, 1988 [1836]), 49 and 48. Humboldt conceived of language as an "organism," the fundamental notion in romantic-idealistic philosophy that served to mediate between nature and *Geist*. The change from this dualistic view to a positivist, monist view of language, generally characteristic of the second half of the nineteenth century—from Hegel to Haeckel, as it were—is exemplified in the work of August Schleicher. See E. Cassirer, *Philosophie der symbolischen Formen, Teil 1. Die Sprache*, 2d ed. (Darmstadt: Wissenschaftliche Buchgesellschaft, 1953), 108ff.; and Richards's chapter in this volume. As the epigraph from Roman Jakobson to this chapter indicates, modern linguistics has gotten beyond such a simplistic view of the naturalness of natural language.

6. Humboldt, *On Language*, 49 (note 5).

7. Ibid., 54 (note 5); W. v. Humboldt, *Schriften zur Sprachphilosophie*, in *Werke*, vol. 3, ed. Andreas Flitner and Klaus Giel (Darmstadt: Wissenschaftliche Buchgesellschaft, 1963), 154.

8. Humboldt, *Schriften zur Sprachphilosophie*, 180–81 (note 7).

9. I use the term *programming language* loosely to include specification languages and (visual) modeling languages. Interfaces that foster a certain kind of interaction are formal languages (or programming languages) too, although they might not appear so to the user. It should be noted that throughout this paper I use the term *interaction* almost exclusively in connection with computers.

10. R. Barthes, "Die Imagination des Zeichens," in R. Barthes, *Literatur oder Geschichte* (Frankfurt am Main: Suhrkamp Verlag, 1969), 36–43.

11. P. E. Agre, "Surveillance and Capture: Two Models of Privacy," *The Information Society* 10 (1994): 101–27, 107. Computer languages very soon came to serve as a model or framework for rational behavior, as for example with Herbert Simon: "Let us first propose a general form for a language of actions, fashioned after modern computer languages" (H. A. Simon, "The Logic of Rational Decision," *British Journal for the Philosophy of Science* 16 [1965]: 169–86, 181). Such "social reverse-engineering" is only one application of a more general epistemological reflection on computer technology as a model for many things. "By a conceptual back-formation characteristic of Western scientific thinking, the division between hardware and software is now being observed in the natural or social world, and has become a new way to express the conflict between ideas of determinism and free will, nature and nurture, or genes and culture. . . . Thus 'software' becomes a viable metaphor for all symbolic activity, apparently divorced from the technical context of the word's origin" (E. Moglen, "Anarchism Triumphant: Free Software and the Death of Copyright," *First Monday* 4, no. 8 [August 1999]: http://www.firstmonday.dk/issues/issue4_8/index.html [27 May 2000]).

12. The term *communicative (inter)action* would be somewhat misleading because Jürgen Habermas has given a special meaning to it. In programming, all action happens within a semiotic field, and communication for acting is at least as important as the activity of communication. By modeling with agency (as is characteristic for this third phase) the two notions finally become indistinguishable. In this sense, the term *performative interaction* refers to John Austin's concept of a "performative act," defined as a speech act that, like a promise, enacts its own content. J. L. Austin, *How to Do Things with Words: The William James Lectures Delivered at Harvard University in 1955*, ed. J. O. Urmson (Oxford: Oxford University Press, 1962). Furthermore, my use of the term *performative interaction* also connotes a subtext of performance understood as assigning roles to heterogeneous actors.

13. I have covered both lines of development in more detail in the perspective of a history of ideas in Jörg Pflüger, "Writing, Building, Growing. Leitvorstellungen der Programmiergeschichte" and "Konversation, Manipulation, Delegation. Zur Ideengeschichte der Interaktivität," in *Geschichten der Informatik. Visionen, Paradigmen und Leitmotive*, ed. H. D. Hellige (Berlin: Springer, forthcoming in 2002). In these articles, as in this chapter, I focus on the discourse of computer science; I have less interest both in the truthfulness of these claims, and in the daily practice of programming, which is governed by completely different conditions such as time pressure and narrow-minded perseverance of routine.

14. F. P. Brooks, "No Silver Bullet: Essence and Accidents of Software Engineering," *IEEE Computer* (April 1987): 10–19, 18.

15. W. Coy, "Automata-Tools-Media: Three Views on Computing," in *Computer Science, Communication & Society: A Technical and Cultural Challenge*, ed. A. Bürgi-Schmelz et al. (Bern: Swiss Informaticians Association & Swiss Sociological Society, 1993).

16. F. L. Bauer and K. Samelson, "The Problem of a Common Language, especially for Scientific Numeral Work (Motives, Restrictions, Aims and Results of the Zurich Conference on Algol)," *Proc. Int'l Conf. Information Processing, UNESCO, Paris 15–20 June 1959* (Paris: UNESCO, 1960): 120–25, 123.

17. A. M. Turing, "Lecture to the London Mathematical Society on 20 February 1947," in *A. M. Turing's ACE Report of 1946 and other Papers*, vol. 10 in the Charles Babbage Institute Reprint Series for the History of Computing, ed. B. E. Carpenter and R. W. Doran (Cambridge, Mass.: MIT Press, 1986), 106–24, 122.

18. The semantics of a programming language simply denotes what a computer computes when fed with a program and data. However, there are different concepts of formal semantics: *Operational semantics* describes algebraically the operations of a mathematical model of a computer that are evoked from corresponding syntactical constructs of the programming language. *Denotational semantics* deals with computational behavior on a more abstract level; complex operations are built from more elementary ones, which are modeled as mathematical functions without regard to how they are realized operationally.

19. Dates are hard to fix precisely in computer science. As every PC-owner has discovered, the announcement of a new computer product does not necessarily mean that it will be released, nor does its release mean that it will function satisfactorily. FORTRAN was first specified in 1954, officially announced in 1956, released in 1957, and only a few months later began to work satisfactorily. The development of FORTRAN illustrates in itself the shift from a mathematical way of talking and thinking toward a notion of language. It's true that, in the original specification, the developers spoke of the FORTRAN language in addition to the FORTRAN system, but the elementary units of such an imperative language, which are later generally called *statements*, "were called *formulas* throughout the 1954 document." In other systems at that time, they were called *equations* or *operations*. D. E. Knuth and L. Trabb-Pardo, "The Early Development of Programming Languages," in *Encyclopedia of Computer Science and Technology* (New York: Marcel Dekker, 1977), 419–93, 461 (reprinted in N. Metropolis, J. Howlett, and G.-C. Rota, eds., *A History of Computing in the Twentieth Century* [New York: Academic Press, 1980], 197–273).

For a detailed study of programming languages in the pre-Babel days, see Knuth and Trabb-Pardo, "Early Development of Programming Languages." For the further history of programming languages, see J. E. Sammet, *Programming Languages: History and Fundamentals* (Englewood Cliffs, N.J.: Prentice Hall, 1969); P. Wegner, "Programming Languages—The First 25 Years," *IEEE Transactions on Computers*, C-25 (December 1976): 1207–25; N. Metropolis et al., eds., *History of Computing*; and R. H. Wexelblat, ed., *History of Programming Languages, Proceedings of the ACM SIGPLAN History of Programming Languages Conference, Los Angeles, California, June 1–3, 1978* (New York: Academic Press, 1981).

20. J. W. Backus et al., "The FORTRAN Automatic Coding System," *Proceedings of the Western Joint Computer Conference* (1957), 1.

21. V. H. Yngve, *Computer Programming with COMIT II* (Cambridge, Mass.: MIT Press, 1972), 2.

22. V. H. Yngve, "COMIT as an IR Language," *Communications of the ACM* (January 1962); reprinted in S. Rosen, ed., *Programming Systems and Languages* (New York: McGraw-Hill, 1967), 375–92, 375 and 379.

23. J. E. Sammet, "The Early History of COBOL," in *History of Programming Languages, Proceedings of the ACM SIGPLAN History of Programming Languages Conference, Los Angeles, California, June 1–3, 1978,* ed. R. H. Wexelblat (New York: Academic Press, 1981), 199–243, 219–20.

24. Humboldt, *On Language,* 61 (note 5).

25. In batch-processing the computing tasks are executed one after another without any user interaction. In the beginning, when programs had to be fed into the computer using tapes or punch cards this was the only mode of operation. In contrast, time-sharing systems that emerged during the early 1960s create the impression of processing many tasks in parallel by allocating small time-slices to each program in turn.

26. G. M. Hopper, "The Education of a Computer," reprinted in *IEEE Annals of the History of Computing* 9 (1988): 271–81, 272 and 273.

27. J. C. R. Licklider, "Man-Computer Symbiosis," *IRE Transactions on Human Factors in Electronic, HFE-1* (March 1960): 4–11; reprinted in *In Memoriam: J. C. R. Licklider 1915–1990* (Palo Alto, Calif.: Digital, Systems Research Center, 1990), 1–15, 2. The [Defense] Advanced Research Projects Agency ([D]ARPA) was created in 1958 in the aftermath of the launch of Sputnik to fund basic research. From 1962 to 1964, Licklider was the head of the newly formed ARPA division called Information Processing Techniques Office (IPTO), where he supported research in time-sharing systems and initiated the development of the ARPANET, the ancestor of the Internet. M. Hauben and R. Hauben, *Netizens: On the History and Impact of Usenet and the Internet* (Los Alamitos, Calif.: IEEE Computer Society Press, 1997).

28. W. D. Orr, "The Culler-Fried Approach to On-Line Computing," in *Conversational Computers,* ed. W. D. Orr (New York: Wiley, 1968), 23–28, 26.

29. D. T. Ross, "Gestalt Programming: A New Concept in Automatic Programming," in *AFIPS Conference Proceedings,* no. 9 (WJCC, 1956): 5–11, 5.

30. Ibid., 6 (note 29).

31. Ibid., 5, 8, and 7 (note 29).

32. Significantly the acronym MAC of the famous time-sharing project at MIT at the beginning of the 1960s had two meanings: "Multiple-Access Computer" and "Machine-Aided Cognition."

33. J. C. R. Licklider, "The Computer in the University," in *Computers and the World of the Future,* ed. M. Greenberger (Cambridge, Mass.: MIT Press, 1962), 203–17, 207, 214, and 216.

34. Ross, "Gestalt Programming," 8 (note 29).

35. Moglen, "Anarchism Triumphant" (note 11).

36. One such means of education used today is the concept of *contracts* in the object-oriented programming language *Eiffel,* which are strictly speaking, modified logical constructs, but serve to teach the programmer to clarify properties of his or

her code. B. Meyer, "From Structured Programming to Object-Oriented Design: The Road to Eiffel," *Structured Programming* 10 (1989): 19–39.

37. D. E. Knuth, "Literate Programming," *Computer Journal* 27 (1984): 97–111, 97.

38. D. T. Ross, "Structured Analysis (SA): A Language for Communicating Ideas," *IEEE Transactions on Software Engineering* ser. 3, no. 1 (1977): 16–34.

39. J. C. R. Licklider, "Potential of Networking for Research and Education," in *Networks for Research and Education: Sharing Computer and Information Resources Nationwide*, ed. M. Greenberger (Cambridge, Mass.: MIT Press, 1974): 44–50, 44.

40. T. O'Reilly, "Lessons from Open-source Software Development," *Communications of the ACM* 42, no. 4 (1999): 33–37, 35. Here the computer technology is surfing the Zeitgeist, since the "blurring of boundaries" between subject and object is also fashionable in postmodern theories, for example the sociology of technology of Bruno Latour, *On Technical Mediation*. The Messenger Lectures on the Evolution of Civilization, Cornell University, April 1993 (Lund University: Institute of Economic Research, School of Economics and Management, Working Paper Series, 9 [1993]); or the popular writings of Sherry Turkle, for example, "Constructions and Reconstructions of Self in Virtual Reality: Playing in the MUDs," *Mind, Culture and Activity* 1 (1994): 158–67; Sherry Turkle, *Life on the Screen: Identity in the Age of the Internet* (London: Phoenix, 1997).

41. L. Wall, "The Origin of the Camel Lot in the Breakdown of the Bilingual Unix," *Communications of the ACM* 42, no. 4 (1999): 40–41, 41. Of course, the idea of a culture based on a programming language is not a new one. Peter Wegner associated with it the obstinate adherence to certain mental attitudes mediated by the languages programmers are used to. "Widely used programming languages like FORTRAN have come to represent a way of thinking to large groups of computer users, and new programming languages with the new way of thinking they represent constitute a challenge to the 'computer culture' that has been built up around established programming languages" (P. Wegner, "Three Computer Cultures: Computer Technology, Computer Mathematics, and Computer Science," *Advances in Computers* 10 [1970]: 7–78, 13).

42. H. H. Goldstine and J. von Neumann, "Planning and Coding Problems for an Electronic Computing Instrument," part II, vol. 1. Report prepared for U.S. Army Ord. Dept. (1947); reprinted in A. H. Taub, ed., *John von Neumann—Collected Works*, vol. 5 (Oxford: Pergamon Press, 1963), 5:80–151, 100.

43. Therefore, it was recommended for a long time that programmers learn an assembly language: "COBOL, RPG, FORTRAN, ALGOL, and PL/1 programmers could do a better job if they know how to program in assembly language. The reason for this is that assembly-language programming puts the programmer closer to the computer, giving the programmer a better knowledge of how the computer works and what affects the efficiency of computer's operation." N. Chapin, *360 Programming in Assembly Language*, 2d ed. (New York: McGraw-Hill, 1968), v.

44. There was a contemporary discussion in linguistics about the different situation of the speaker and the listener and the necessity of modeling their use of language with different kinds of grammars. "For the receiver the message presents many ambiguities which were unequivocal for the sender" (Jakobson, "Linguistics and Communication Theory," 249 [note 4]). "For the hearer, then, a grammatical system must be viewed as a stochastic process" (C. F. Hockett, "Grammar for the Hearer," in *Proceedings of the Twelfth Symposium in Applied Mathematics: Structure of Language and its Mathematical Aspects*, ed. R. Jakobson [Providence, R.I.: American Mathematical Society, 1961], 220–36, 220). However, in contrast to the human listener, it was mandatory in computer science that the mechanical reader be able to decode unambiguously.

45. A *procedure* is a parametrized piece of program text in procedural languages that can be "called" from different places in the main program to compute some functionality with various data sets.

46. R. K. Moore and W. Main, "Interactive languages: design criteria and a proposal," *AFIPS Conference Proceedings FJCC* 33 (1968): 193–200, 194.

47. J. C. Shaw, "JOSS: Experience with an Experimental Computing Service for Users at Remote Consoles," RAND Report P-3149; reprinted in W. D. Orr, ed., *Conversational Computers* (New York: Wiley, 1968), 15–22, 17 and 16.

48. This impression may also have been enhanced by the social experiences of a community of researchers working with time-sharing systems on the same problems and using the same (sometimes formal) language.

49. Shaw, "JOSS," 18 (note 47).

50. S. L. Marks, "JOSS—Conversational Computing for the Nonprogrammer," *Annals of the History of Computing* 4, no. 1 (1982): 35–51, 51.

51. L. E. S. Green, E. C. Berkeley, and C. C. Gotlieb, "Conversation with a Computer," *COMPUTERS and AUTOMATION* (October 1959): 9–11; P. J. McGovern, "Computer Conversation Compared with Human Conversation," *COMPUTERS and AUTOMATION* (September 1960): 6–11. McGovern, with the Turing Test firmly in mind, concluded: "Prediction 1: It will not be many years—I would estimate hardly more than ten years—before the operating of computers with ideas will be widespread. Prediction 2: When computers do operate generally with ideas, it will be impossible for a human being in another room to tell whether he is conversing with a computer or with a human being" (11).

52. L. Mezei, "Conversation with a Computer," *Datamation* (January 1967): 57–58; Barry W. Boehm, "The Professor and the Computer: 1985," *Datamation* 13, no. 8 (1967): 56–58.

53. Whereas in batch-processing the computer functions as an independent automaton, the early concept of interactivity conceived the coupling of human and machine as a symbiotic automaton. In the phantasmagoria of an "on-line thinking apparatus" cybernetic conceptions survived that had otherwise been supplanted by computer science. "In the on-line mode, human intelligence and computing machinery interact as elements in a continuous feedback loop." "When a problem solver is

working on-line with computing machinery, his personal intelligence becomes a integral part of the process. Mind and machine interact continuously and simultaneously in a coordinated attack upon the problem at hand." "The concept underlying on-line problem solving is control through feedback, which was also the central idea of Norbert Wiener's *cybernetics*" (Orr, "Culler-Fried Approach," 23 and 25 [note 28]).

54. In the conversation mode it is again possible to think of a personal relationship between the human and his or her machine. Contrary to the earlier machines, education no longer relies on commands, but on the adaptation of the machine to the idiosyncrasies of its user. In 1970, Nicholas Negroponte still dreamed of a "progressively intimate association of the two dissimilar species in the symbiosis": "Imagine a machine that can follow your design methodology and at the same time discern and assimilate your conversational idiosyncrasies. This same machine, after observing your behavior, could build a predictive model of your conversational performance. Such a machine could then reinforce the dialogue by using the predictive model to respond to you in a manner that is in rhythm with your personal behavior and conversational idiosyncrasies." N. Negroponte, *The Architecture Machine: Towards a More Human Environment* (Cambridge, Mass.: MIT Press, 1970); cited in R. M. Baecker and W. A. S. Buxton, eds., *Readings in Human-Computer Interaction: A Multidisciplinary Approach* (San Mateo, Calif.: Morgan Kaufmann Publishers, 1987), 51.

55. M. V. Wilkes, "Computers Then and Now," *Journal of the ACM* 15, no. 1 (1968): 1–7, 4.

56. H. D. Bennington, "Production of Large Computer Programs," *IEEE Annals of the History of Computing* 5, no. 4 (1983): 350–61, 352.

57. F. T. Baker and H. D. Mills, "Chief Programmer Teams," *Datamation* 19, no. 12 (1973): 58–61, 58.

58. This is the title of Alan Kay's dissertation, which had a strong influence on the development of the *Dynabook*, a conceptual precursor of personal computers and on the concept of manipulation-interfaces as well (A. Kay, "The Reactive Engine," Ph.D. thesis, University of Utah, 1969). The spirit of action let the developers of the *Dynabook* exclaim enthusiastically: "Imagine having your own self-contained knowledge manipulator in a portable package the size and shape of an ordinary notebook." *Personal Dynamic Media* (Xerox Corporation Palo Alto Research Center, Learning Research Group, March, 1975), 2.

59. D. H. H. Ingalls, "Design Principles Behind Smalltalk," *BYTE* 6, no. 8 (August 1981): 286–98, 286.

60. The operating system UNIX had a tool concept, too, with special functions to be flexibly combined to form new ones. Thus the user has access to a toolbox, but the building of his or her own tools follows a linguistic logic and not (yet) the model of manipulation: complex tools are "expressively" combined, in a way that compound expressions are composed out of simple ones in language. Not surprisingly, UNIX was soon enhanced by graphical interfaces.

61. J. Johnson et al., "The Xerox Star: A Retrospective," *Computer* (September 1989): 11–26, 15.

62. Smith et al., "Designing the Star User Interface," *BYTE* 7, no. 4 (1982): 242–82, 656.

63. A. Kay, "User Interface: A Personal View," in *The Art of Human-Computer Interface Design*, ed. B. Laurel (Reading, Mass.: Addison-Wesley, 1990), 191–208, 196.

64. Johnson et al., "Xerox Star," 15 (note 61).

65. Kay, "User Interface," 201 (note 63).

66. Xerox, *Personal Dynamic Media*, 6 (note 58).

67. Kay, "User Interface," 201 (note 63).

68. K. Nygaard, "Program Development as a Social Activity," in *Information Processing 86*, ed. H.-J. Kugler (Amsterdam: Elsevier Science Pub., 1986), 189–98.

69. Though, or indeed because, the influence of language on the varying conceptions of "building" is hardly evident, I would like to claim that the manner and method of modeling can be well described with generalized rhetorical operations. I use the categories *metaphor* and *metonymy* as more or less equivalent with the terms *paradigmatic* and *syntagmatic* according to the "two-axes model" of Roman Jakobson, describing semiotic operations of selection and combination respectively. See C. Lévi-Strauss, *The Savage Mind* (Chicago: University of Chicago Press, 1966); and R. Jakobson and M. Halle, *Fundamentals of Language* (The Hague: Mouton, 1956). A metaphorical relation means here the affirmation of similarity between different concepts or different model levels; metonymies grasp their inner relations of contiguity or the syntagmatic relations of signs. Extended to technical artifacts, one can speak of a specification as a formal metaphor and of its combinatorial exploration as a metonymical process. See J. Pflüger, "Distributed Intelligence Agencies," in *HyperKult*, ed. M. Warnke, W. Coy, and G. C. Tholen (Basel: Stroemfeld, 1997), 433–60, 441–42. The framework of rhetoric not only yields analogies but also provides an analytical instrument for a contrastive study of approaches in computer science and other media, which I will exploit further in future papers.

70. In the rhetorical tradition, around the same time, pragmatic-oriented stylistics was intended similarly to overcome the traditional concept of "style as dress," emphasizing the act of putting into words as an integral moment of creation, not merely the ornamental elocutionary sequel of cognitive invention. See, for example, G. Antos, *Grundlagen einer Theorie des Formulierens. Textherstellung in geschriebener und gesprochener Sprache* (Tübingen: Niemeyer, 1982), where Gerd Antos considers the production of text as a problem-solving process.

71. H. Stoyan, *Programmiermethoden der Künstlichen Intelligenz 1/2* (Berlin: Springer, 1988), 184.

72. Meyer, "From Structured Programming to Object-Oriented Design," 22 (note 36).

73. Accordingly, for object-oriented languages there is no convincing (theoretical) concept of formal semantics, and hardly anybody has tried to come up with one.

74. P. Kearney, "Personal Agents: A Walk on the Client Side," in *Agent Technol-*

ogy. Foundations, Applications, and Markets, ed. N. R. Jennings and M. J. Wooldridge (Berlin: Springer, 1998), 125–38, 134.

75. Success stories of the open-source and the free-software movement like the Linux project are prototypical examples of such a decentralized bottom-up development process, growing software as a cooperative effort of individuals and groups distributed all over the world, even if they use more traditional programming techniques.

76. Kearney, "Personal Agents," 126 (note 74).

77. Agre, "What are Plans for?," 25 (note 101).

78. B. Laurel, *Computers as Theatre* (New York: Addison-Wesley, 1991).

79. Kay, "User Interface," 198 (note 63).

80. Xerox, *Personal Dynamic Media*, 48 (note 58).

81. P. Maes, "Agents that Reduce Work and Information Overload," *Communications of the ACM* 37, no. 7 (1994): 30–40, 31.

82. B. Laurel, "Interface Agents: Metaphors with Character," in *The Art of Human-Computer Interface Design*, ed. B. Laurel (New York: Addison-Wesley, 1990), 355–65, 360.

83. Laurel, "Interface Agents," 358 (note 82).

84. Kay, "User Interface," 205 (note 63).

85. L. N. Foner, "What's An Agent, Anyway? A Sociological Case Study," (1993) http://foner.www.media.mit.edu/people/foner/Julia (27 May 2000).

86. Of course there are always exceptions. Interestingly, what was probably the first programming language, Konrad Zuse's *Plankalkül* from 1945, had a rather elaborated data concept. The *Plankalkül*, which only existed on paper, was in many aspects far ahead of its time, but it had no practical influence on the further development of programming languages. Most of the languages developed in the ensuing years started "at the other end, by asking what was possible to implement rather than what was possible to write" (Knuth and Trabb-Pardo, "Early Development of Programming Languages," 428 [note 19]). Later, the commercially oriented language COBOL emphasized data structures rather than algorithmic refinement.

The procedural (or equally functional) view is not necessarily linked to the existence of language constructs such as procedures or functions. (Along with procedural languages came functional programming languages that reproduce mathematical functions even more clearly.) Characteristic of the procedural approach is that it models problem-solving processes, thus subordinating everything to algorithms. In assembly languages and primitive imperative languages the organization of computation is represented directly and functional units are not yet packed in separate program constructs—procedures or functions, meaning the algorithm is not yet dissected into its logical parts. The unstructured algorithmic representation in the early years corresponded well to the conception of "writing" a program, as the algorithm constitutes the "narrative flow of action." Similarly, interactive programming, which can be regarded as "algorithmization in progress," shows no structure and is governed by the turn-taking "flow of words."

87. In his attempt to design a universal "philosophical language," John Wilkins

intended to do away with verbs altogether (J. Wilkins, *An Essay Towards a Real Character and a Philosophical Language* [London, 1668], 298 and 303). See also Geneva's chapter in this volume.

88. M. Kline, *Mathematics in Western Culture* (New York: Oxford University Press, 1953), 274.

89. "Phase I Report—Language Structure Group of the CODASYL Development Commitee, An Information Algebra," *Communications of the ACM* 5 (1962): 190–204, 191 and 190.

90. Modeling with so-called entity-relationship diagrams became popular with respect to data base organization in the 1970s. Not surprisingly, a data-centered view of programming developed earlier in connection with data bases than within the realm of programming language design. "The data-centered point of view considers the data structure (data base) as the central part of a problem specification and views programs as 'bugs' which crawl around the data base and occasionally query, update or augment the portion of the data base at which they currently reside. . . . The data-centered view of programming led in the 1970's to the development of data-base languages and systems" (P. Wegner, "Programming Languages," 1223 [note 19]).

91. Wilkes, "Computers Then and Now," 4 (note 55).

92. This type of modeling abstracts in the sense of the Platonic idea, capturing the essence of a number of similar phenomena. There are related programming languages, so-called prototyping languages, which do not distinguish between object and class. They support an even more experimental approach to modeling: one defines a prototype on encountering something new and later makes copies or modifications of it.

93. The object-manipulating user technique in the interface proceeds analogously with object-oriented programming technique. The usage of application programs, which are invoked and then fed with data, corresponds to the procedural view. Under the desktop metaphor, however, the object of work has priority over the means of labor: one does not call an editor, but opens a document. First an object is selected, for which then an operation is activated, whereas with the dialog languages, first the operation is chosen, for which then the parameters and data have to be given. The developers of the Xerox Star formulated the inversion of the functional-oriented procedure with the same linguistic metaphor: "Commands in Star take the form of noun-verb. You specify the object of interest (the 'noun') and then invoke a command to manipulate it (the 'verb')" (Smith et al., "Star User Interface," 660 [note 62]). Alan Kay points out the common root of developing and application: "In both cases we have the *object* first and the *desire* second. This unifies the concrete with the abstract in a highly satisfying way" (Kay, "User Interface," 197 [note 63]). In contrast, Thomas Erickson views in this "simple noun-verb syntax" a form of pidgins, owed the novelty and strangeness of interactivity, which always arises when two "radically different cultures" meet for the first time. T. D. Erickson, "Interface and the Evolution of Pidgins: Creative Design for the Analytically Inclined," in *The Art of Human-Computer Interface Design*, ed. B. Laurel (New York: Addison-Wesley, 1990), 11–16, 12ff.

94. G. Kiss, J. Domingue, and C. Hopkins, "A Multi-Agent System for Telephone Network Management," http://kmi.open.ac.uk/kmi-abstracts/hcrl-tr-72-abstract.html, 1991.

95. N. R. Jennings and M. J. Wooldridge, "Applications of Intelligent Agents," in *Agent Technology. Foundations, Applications, and Markets*, ed. N. R. Jennings and M. J. Wooldridge (Berlin: Springer, 1998), 3–28, 4.

96. Whereas one spoke with respect to object-orientation more vaguely of a "community of communicating objects," it is now claimed for agent-oriented design that one has to deal with "a new programming paradigm, based on a societal view of computation" (Y. Shoham, "Agent-oriented Programming," *Artificial Intelligence* 60 [1993]: 51–92, 51). Agent societies can be seen as a technological reflection of a service society, based on both delegation and cooperation as the dominant model of subjectivity and "social" interaction.

97. I. Greif, "Desktop Agents in Group-enabled Products," *Communications of the ACM* 37, no. 7 (1994): 100–105, 104.

98. Jennings and Wooldridge, "Applications of Intelligent Agents," 7 (note 95). Since the 1970s, there has been a shift of perspective in Artificial Intelligence from the individual to a social "intelligence" akin to such distributed modular systems. This shows in the change of metaphor for a programmed "problem solver." Whereas it was first conceived of as a "simple personality" (A. Newell, *Some Problems of Basic Organization in Problem-Solving Programs*, Rand Corporation Memorandum RM-3283-PR [December 1962]; cited in C. Hewitt, "Viewing Control Structures as Patterns of Passing Messages," *Artificial Intelligence* 8 [1977]: 323–64, 323) it is now imagined as a "society of experts." C. Hewitt, "Viewing Control Structures as Patterns of Passing Messages," 324.

99. D. Wenger and A. R. Probst, "Adding Value with Intelligent Agents in Financial Services," in *Agent Technology. Foundations, Applications, and Markets*, ed. N. R. Jennings and M. J. Wooldridge (Berlin: Springer, 1998), 303–25, 304. Today one talks again of partnership: "A fundamental rethink is needed about the nature of interaction between computer and user. It must become an equal partnership—the machine should not just act as a dump receptor of task description, but should cooperate with the user to achieve their goal" (Jennings and Wooldridge, "Applications of Intelligent Agents," 7 [note 95]). However, this is a very different conception from former "discourse circles." Now we have very different kinds of agents acting autonomously, which live in an "actor-network" and encounter each other as more or less equals.

100. Kearney, "Personal Agents," 127 (note 74).

101. Of course, there are risks connected with this approach, and the emergent behavior of an agent-based system does not suit for areas where security is a concern. However, we have already seen that in the name of interactivity people are willing to give up control: "Activity arises through interaction, not through control" (P. E. Agre and D. Chapman, "What Are Plans for?" in *Designing Autonomous Agents: Theory and Practice from Biology to Engineering and Back*, ed. P. Maes [Cambridge, Mass.: MIT Press, 1990], 17–34, 32).

102. Surveying the wide field of computer-mediated communication, it appears that computer science has largely subscribed to a pragmatics dealing with technical preconditions for communicative action (in the sense of Jürgen Habermas). In modern computer science, questions of platform independence, of transmission protocols, and of the specification of rules for communication are much more vital than the underlying problems of computation. For example, the http protocol regulates how to communicate in the World Wide Web, and the P3P protocol governs the control of privacy data. These pragmatic efforts could be apostrophized as a search for technical dispositives for a virtual "ideal speech situation." In the technological sphere such a kind of "universal pragmatics" will, however, exhibit no generalized principles, but only a lot of protocols. See K.-O. Apel, ed., *Sprachpragmatik und Philosophie* (Frankfurt am Main: Suhrkamp, 1976); or J. Habermas, *Theorie des kommunikativen Handelns, I–II* (Frankfurt am Main: Suhrkamp, 1981).

103. Humboldt, "On Language," 62 (note 5).

104. I cannot discuss here the manifold repercussions of computer science, even with respect to language, but I would like to illustrate the reciprocal impact of the language analogy with an example that demonstrates how an analogy of a language analogy sets a new reality: computer science proceeds according to a theory of correspondence in order to model work flow in organizations and other activities. However, its representation schemes have the decisive effect of (re)structuring human activities in analogy with a formal language. "They each employ formal 'languages' for representing human activities. Human activity is thus effectively treated as a kind of language itself, for which a good representation scheme provides an accurate grammar. This grammar specifies a set of unitary actions—the 'words' or 'lexical items' of action. . . . It also specifies certain means by which actions might be compounded—most commonly by arranging them into a sequence, although various languages provide more sophisticated means of combination (for example, conditional and iterated sequences)" (Agre, "Surveillance and Capture," 108–109 [note 11]). Phillip Agre describes quite different activities that are regulated by such a "grammar of action." In contrast to the rigid organization of work in automation or Taylorism a grammar of action constitutes a peculiar relation between local freedom and global control. Singular actions can be chosen relatively freely out of a given pool, but in order to conform with others they have to follow the rules of their composition. That freedom of a controlled choice "is precisely the type ascribed to language users by linguistic theories of grammar—for example in Chomsky's . . . notion of 'free creation within a system of rule.'" (Agre, "Surveillance and Capture," 117 [note 11]).

105. B. Laurel, *Computers as Theatre*, 1 (note 78).

106. The entanglement of user and agent in a common language even in private issues may be illustrated by the description of the P3P protocol given by the lawyer Lawrence Lessig: "P3P is a design that would enable individuals to select their preferences about the exchange of private information, and then enable agents to negotiate the trade of such data when an individual connects to a given site. . . . P3P functions as a language for expressing preferences about data, and as a framework

within which negotiations about those preferences could be facilitated" (L. Lessig, "The Law of the Horse: What Cyberlaw Might Teach," *Harvard Law Review* 113, no. 2 [1999]: 501–49, 524).

107. "Edward A. Feigenbaum: The Power of Knowledge," in *Out of Their Minds: The Lives and Discoveries of 15 Great Computer Scientists*, ed. D. E. Shasha and C. A. Lazere (New York: Copernicus, 1995), 209–22, 222.

108. L. G. Terveen and L. T. Murray, "Helping Users Program Their Personal Agents," *CHI* 96 (1996): 355–61, 355.

INDEX

Writing Science

Felicia McCarren, *Dance Pathologies: Performance, Poetics, Medicine*

Timothy Lenoir, ed., *Inscribing Science: Scientific Texts and the Materiality of Communication*

Niklas Luhmann, *Observations on Modernity*

Dianne F. Sadoff, *Sciences of the Flesh: Representing Body and Subject in Psychoanalysis*

Flora Süssekind, *Cinematograph of Words: Literature, Technique, and Modernization in Brazil*

Timothy Lenoir, *Instituting Science: The Cultural Production of Scientific Disciplines*

Klaus Hentschel, *The Einstein Tower: An Intertexture of Dynamic Construction, Relativity Theory, and Astronomy*

Richard Doyle, *On Beyond Living: Rhetorical Transformations of the Life Sciences*

Hans-Jörg Rheinberger, *Toward a History of Epistemic Things: Synthesizing Proteins in the Test Tube*

Nicolas Rasmussen, *Picture Control: The Electron Microscope and the Transformation of Biology in America, 1940–1960*

Helmut Müller-Sievers, *Self-Generation: Biology, Philosophy, and Literature Around 1800*

Karen Newman, *Fetal Positions: Individualism, Science, Visuality*

Peter Galison and David J. Stump, eds., *The Disunity of Science: Boundaries, Contexts, and Power*

Niklas Luhmann, *Social Systems*

Hans Ulrich Gumbrecht and K. Ludwig Pfeiffer, eds., *Materialities of Communication*

Printed and bound by CPI Group (UK) Ltd, Croydon, CR0 4YY

16/04/2025

14658401-0004